IS
THE
MOON
UPSIDE
DOWN?

IS THE MOON UPSIDE DOWN?

A Quicke Guide to the Cosmos

GREG QUICKE

EBURY
PRESS

EBURY PRESS

UK | USA | Canada | Ireland | Australia
India | New Zealand | South Africa | China

Ebury Press is part of the Penguin Random House group of companies whose
addresses can be found at global.penguinrandomhouse.com

Penguin
Random House
Australia

First published by Ebury Press in 2020

Cover images by Shutterstock
Cover design by Adam Laszczuk © Penguin Random House Australia Pty Ltd
Typeset in 12.5/17.5 pt Berkeley by Midland Typesetters, Australia

Printed and bound in Australia by Griffin Press, part of Ovato, an accredited
ISO AS/NZS 14001 Environmental Management Systems printer

 A catalogue record for this
book is available from the
National Library of Australia

ISBN 978 1 76089 505 1

penguin.com.au

*Dedicated to all of you with
the power to think for yourselves,
to observe and to let your heart
guide you the right way through life.*

Contents

What a Spin!

How fast are you going right now? Are you sitting still? Are you riding on a bus, a train, or cruising along in a car? Maybe you're on a plane. Whatever you guess your current speed to be, I think it might fall a little short.

If you're somewhere near the equator, at this very moment you're zipping along at around 1670 km/h. To put that into perspective, you're racing from east to west at four times the top speed of a MotoGP motorcycle going down the front straight at Phillip Island.

This is the rotational speed of planet Earth if you're at the equator. Picture that racing bike screaming past in a blur. Can you believe you're travelling that quick right now? Can you feel it?

It's more likely, though, that you are north or south of the equator, where most of us live. The further from zero degrees latitude (aka the equator) and the closer to the poles you go, the smaller the circle you'll be travelling in the same 24-hour period. With not as far to go in the same amount of time, you'll be going slower than you would be at the equator.

In my hometown of Broome on Western Australia's wild Kimberley Coast – 18 degrees south of the equator – I'm motoring around in an easterly direction at around 1333 km/h, or 370 m per second. Life in the tropics is pretty fast.

People in Perth, Melbourne or Sydney are doing a bit under 1100 km/h, as are the residents of Los Angeles and Tokyo, since they're all similar distances above and below the equator. New Yorkers are moving at more like 900 km/h, while in London – way up at 52 degrees north – they're dawdling along at 700 km/h.

However, if you're trudging across the ice 30 km from one of the poles you're doing a much more sedate 4 km/h, and if you're standing at the North or South Pole you're simply spinning around on the spot – once a day.

Yet as you go about your terrestrial life – perhaps sitting and reading this book, riding on a bus or walking up the street to get a coffee – I'd be willing to bet you rarely consider the mesmerising fact that we are all part of a never-ending, high-speed planetary spin-cycle.

But that's not even the half of it.

In the minute or so you've spent reading this far, Earth has dragged you at least a couple of thousand km in another direction. Right now, the planet is blasting you through space at about 107,000 km/h as it careens around the Sun. So not only are you spinning from west to east at anywhere up to 1673 km/h, you're hurtling more than 100 times faster than a jumbo jet to a point on the other side of our nearest star and back again on a repeat round trip, which carries you in the order of 940 million km every year. How are you feeling?

How far will you travel in your lifetime? If you live to be 80, you'll have covered about 75.4 billion km – give or take a bus ride or two.

Like everyone else born in 1961, I've already travelled more than 54 billion km on this fantastic voyage through space, but, unlike most people, I *can* feel this movement. Not because I have special powers or I'm extra perceptive, but simply because I've been watching the movements of the cosmos for so long I've tuned into the rhythm of the great celestial dance we're all a part of. It's not that hard; all you have to do is watch.

Like many of you, I was mesmerised and amazed by the stars as a child. As an adult I still sleep under the outback Australian skies so often that we get to know each other more and more all the time. Early on, intrigue bloomed into total fascination and, ultimately, my love of the cosmos led to my life's calling as an astronomer.

Over the past 25 years I've shared my observations of the night sky with more than 100,000 people at my outdoor observatory, Astro Tours, smudged into the red soil just outside of Broome. More recently I've been lucky enough to widen that audience to millions around the world, thanks to invitations to appear on the BBC series *Stargazing Live* and the Australian version of the show on ABC TV, where I play the practical astronomer to Professor Brian Cox's theoretical particle physicist. The exposure led to me presenting my own 10-part series on ABC TV, *A Stargazer's Guide to the Cosmos*.

Nowadays – thanks to some cheeky lark on Twitter – I seem to be stuck with the nickname 'Space Gandalf'. While I might share a hairstyle with Tolkien's great wizard, I'm just an ordinary bloke – a self-taught bush mechanic and former pearl diver, a motorbike nut and a mad surfer, who has watched in awe as our planet spins and tumbles through the ever-changing universe.

By the way, I realise you already know that the Earth turns. We learned it in school, we read about it in books and you've probably seen shows about it on television. There have been some good ones on ABC TV! I'm also guessing, however, that you've rarely – if ever – pondered the magnitude of the forces at play around us.

So, if you're up for finding out more and you'd like a tour guide as you come to grips with the most amazing yet funda-mental journey of your life, keep reading. I'll do my best to tune you in to the stars. They've changed my life and I reckon there's a good chance they'll do the same for you.

Chapter 1

Look Straight Up

Our launch pad for the exploration of the cosmos is our very own exquisite little planet, Earth. It is an amazing travelling spaceship fitted out with countless extraordinary and useful features. Some of my favourites include Earth's mighty trees, her curling liquid waves, her lavish gardens of fresh food, the oceans filled with life, clean air, sky-scraping mountains, terrific people and awesome motorcycles.

You find life everywhere you look on Earth: at the peaks of the highest mountains and at the bottom of the deepest ocean trenches. Organisms live in the coldest ice sheets and can be found in the hottest boiling springs. We see life scattered across the floors of the driest deserts and high amongst the rafters of the upper atmosphere.

Much has been said about how the Earth's perfect positioning in relation to the Sun gifts us just the right amount of heat and light for water to be liquid. Our atmosphere offers us the perfect amount of protection from harmful solar rays, while a magnetic field also stands sentry over us as it deflects charged particles that would otherwise fry our DNA.

Some say all of this is a perfect coincidence, others believe it's part of a grander plan, while there are plenty of people who likely haven't thought much about it at all. Whatever your opinion, it doesn't really matter to Earth – she simply is the way she is.

We know our planet has a few bumps that we call mountains and the oceans fill up some pretty deep holes too. Due to the centrifugal force generated by its rotation, Earth bulges at the equator and is flattened out at the poles. An eight-year-old visitor to Astro Tours once informed me that the Earth is an 'oblate spheroid'. For the purposes of getting to know the stars and our relationship to them, however, it's good enough for us to say the Earth is round.

We live on a rocky planet, and a relatively smallish one at 12,756 km in diameter. The solid crust varies from as little as 5 km thick in the ocean beds to as much as 50 km deep on the continents. The crust overlies a semi-liquid mantle of activity that plunges 3000 km down. Inside the mantle, liquid rock in the planet's outer core flows around a solid iron core that is nearly 2500 km in diameter. Even though we haven't seen any of this, it is our current model.

Here on the surface 71% of the crust is wrapped in saltwater oceans, above which floats a 100 km thick blanket of gases that make up our atmosphere. If Earth were the size of a cricket ball, the atmosphere would be as thick as the red paint on the leather. This amazing invisible skin is made of 78% nitrogen, 21% oxygen, plus small amounts of argon, carbon dioxide and other trace gases. It regulates temperatures and, with its H_2O

content, the atmosphere distributes water all over the planet through evaporation and rain.

One of my favourite things about Earth's atmosphere, however, is its transparency. This allows us the extravagant privilege of being able to peer out into the seemingly endless space that envelopes us and observe the majesty of the universe and the mind-stretching astral ballet therein.

I reckon just about everyone remembers staring at the stars as a kid, especially here in Australia. My own early memories of the night skies revolve around family caravanning trips in the 1970s. We'd head off from southern Western Australia to explore the glorious north-west. Away from town and the curse of light pollution, the sky unfurled overhead in a breathtaking sweep from horizon to horizon.

Rather than sleep in the caravan with Mum, Dad and my sister, my little brother and I would shape beds in the big piles of gravel by the side of the road. We'd climb into our sleeping bags, lie on our backs and look straight up. Glued to the ground by gravity as my eyes flitted from constellation to constellation, I really felt that I was lying on the curved surface of the Earth while she turned me through space.

On other nights back home in Pemberton, Dad would take us marron fishing along lovely, lonely stretches of the Warren River. We'd sit on the fallen trees that bridged the banks and wait for the black-clawed marron to wander into our nets dangling in the waters below. Above us a ribbon of sky flanked by the canopies of 60 m karri trees lit up with a million distant suns. I don't know if I knew they were suns at the time; I just knew the stars were beautiful and that I loved them.

Throughout childhood and right through my school years, however, I never dreamed I'd become an astronomer, let alone one who'd wind up telling stories about the universe on TV. I attended the state school in Pemberton, and while I was usually pretty easy to get along with, if things didn't make sense I wanted to keep asking the difficult questions. I now know that teachers often don't know the answers, although at the time I challenged some of them pretty hard.

Although I had some excellent teachers who nurtured me, I was also saddled with a couple whom, in my immaturity, I could have been kinder to. For my troubles, I spent a bit of time in the principal's office, where we got to be quite good friends to this day.

What bugged me most was when an educator clearly didn't know what he or she was talking about.

I once asked a science teacher, 'Exactly what is a flame? What is fire?'

He rattled off some non-answer that avoided the specific question, so I asked again, 'No, no. What *exactly* is the flame itself?'

He still didn't answer, which told me he didn't know. Years later I found the answer myself – flame is plasma. I also learned that space is plasma too. We'll talk about plasma in a later chapter.

On another occasion, during a lesson on trigonometry, I asked my maths teacher, 'Sir, what is this trigonometry used for exactly?'

'Shut up, Quicke!' he barked. 'Just shut up and learn it!'

Once again this let me know my maths teacher didn't have

the foggiest idea what earthly purpose trigonometry served. If he had, he might have told me, 'We use it to measure the distances to the stars,' and I may have done a lot better at trigonometry.

When it came time in high school to nominate a career path, I was pretty clueless. I had, however, just watched a documentary that featured a marine biologist diving on the Great Barrier Reef.

'I want to do that!' I blurted out to the careers advisor. I loved scuba diving – I'd been doing it from the age of 14 – and I figured any career that involved diving was worth pursuing.

'To do that, you'll need to learn marine biology,' the advisor replied.

'That's settled, then,' I said. 'I'll be a marine biologist.'

I left high school with good enough marks to study marine biology at James Cook University in North Queensland. So I arrived in Townsville ready to go diving on the Great Barrier Reef and learn about marine biology – after all, that's what the job looked like when I'd seen it on TV.

I discovered instead that marine biologists spend most of their time peering into microscopes and memorising endless funny names for things that already had ordinary names. While I found this interesting for a while, it wasn't long before the novelty wore off. I quit without ever sitting an exam. Or going diving even once!

I wandered around Australia for a while and in 1982 at the age of 20 I wound up in Broome on WA's Kimberley Coast. It took me a week to realise I was in the world's pearling capital, and when that penny dropped I walked into Bill Reid's Broome Pearls shed in Chinatown.

'Do you guys need any divers?' I asked.

'Start tomorrow,' Bill replied.

I didn't realise it at the time, but this new job marked the beginning of my astronomical career. Instead of working Monday to Friday and having the weekends off, pearl divers work according to the cycle of the moon and its regulation of the tides.

We'd dive for pearl shell 10 days straight over the 'neap tides' and have four days off during 'spring tides'. The movement of the tides in Broome is among the biggest in the world, with up to 10 m between high and low tide. The impact this had on people like me, who fossicked around on the floor of the Indian Ocean for a living, was significant.

The enormous surge of water caused by the massive spring tides creates strong currents that make underwater working conditions difficult and dangerous. Sediments are stirred up, which cuts visibility to almost zero. Meanwhile, the power of the currents is so strong that if you don't hang onto a fixed point you'll be swept away. This is why pearl divers go to the pub during the four days of spring tides.

The neap tides return for the 10 days centred around the first quarter moon and again at the last quarter moon. These are much more gentle for the working diver. The tides don't come in as far or go out as far, which means the currents are slower, the sediments settle, visibility improves and diving conditions are far more pleasant.

These Kimberley mega tides also made it easy for me to see the very obvious influence the moon has on the surface of our planet. The big spring tides happen when the Earth, Sun and moon line up with each other in space. These alignments occur twice a month. You know them as 'full moon', when the moon is opposite the Sun, and 'new moon', when the moon comes between Earth and the Sun.

However, when the Sun and the moon are square to each other at what we call 'first quarter moon' and at 'last quarter moon', the tides slow down, and grow smaller and more manageable – the blessed neap tides.

Diving for the Pinctada maxima pearl shells in the warm, sapphire-coloured waters of Broome's Roebuck Bay, I learned to watch the moon carefully and I paid attention to its position in relation to the Sun. I became attuned to the three-way dance between Earth, the Sun and the moon, as they conjured tides by pulling oceans this way and that according to their orientation at a given time.

After a while I could tell at a glance what time high tide would be, and by watching the moon's phase I could work out how high the tide would be too. (I'll talk more about the tides and the moon cycle in a later chapter.)

Although it was dangerous work, I loved it. I dived throughout the 1982 season and got friendly with sharks, sea snakes and deadly stonefish along the way. The biggest danger we faced, however, was running out of the compressed air that was pumped down a hose from a boat 18 m above us as we worked the ocean floor. When that happened, we had to carefully and slowly make our way to the surface – without giving in to panic – in what's called a 'free ascent'. To do otherwise was to risk a terrible end.

A free ascent is a safe enough procedure, although every time I performed one over that year it showed me very clearly the line between life and death. Having lungs full of compressed air at depth means the same air at the surface expands to the point where it will burst your lungs if you don't blow it out. No fun at all.

The key to a successful free ascent is to blow air out of your lungs as you slowly head for the surface. This goes against every instinct you have to get to the surface ASAP. In theory, you blow air out at about the same speed as if you were whistling and then follow your smallest bubbles up to the surface. I usually blew air out faster than suggested and was always keen to get to the surface sooner rather than later.

Then there were the big phantom whooshes of water over our heads as we filled our baskets with pearl-bearing oysters. Due to the limited visibility in Roebuck Bay, we never saw what caused these weird disturbances, but the waters of the Kimberley Coast are home to all manner of hungry critters, including bull sharks, tiger sharks and even saltwater crocodiles.

I got spooked a few times and eventually had trouble shaking the feeling there was a big finger tapping me on the shoulder, saying, 'I can take you out any time I want.' After diving for one season I decided to quit the pearling industry while I was still ahead – and still alive.

In addition to cash for putting fuel in my motorcycle, pearl diving gave me a very real sense that I was on a planet called Earth. A significant little planetoid world that we call the moon entered my life at that time too. The first celestial cycle I seriously and consciously considered was that of the moon going around the Earth. This is a cycle we all live by, yet one that is below the conscious level for most people in the modern world of cities, smart phones and television.

Tuning in to the way these two worlds play with each other tweaked something deep inside me. I began to recognise that we live our lives in cycles and that there was much, *much* more for me to learn.

Chapter 2

The University of Life

Whenever people ask which university I studied astronomy at, I usually point at the sky. You simply can't learn what I have by sitting in a classroom. I know this because university students come to me to learn the sky. Astronomy gives you the practical tools for watching the very passage of time through the passage of the stars and the planets through space. Watch them for long enough and they'll show you their secrets.

Not only did I not study astronomy at uni, I didn't race out one day and buy an expensive telescope to start looking out into space either. That came much later. What I *did* have at my disposal was a big black Z1000 Kawasaki motorbike and a swag – two effective tools that helped me begin to learn about the cosmos.

With marine biology and pearl diving off the table as long-term career options, I rode across every dirt road from Broome to the Northern Territory with a couple of mates in the wet season of 1983. Every night we'd sleep under the stars in our swags.

My swag is a simple 8 foot by 8 foot piece of waterproof canvas, a two-inch foam mattress, sheets and a blanket, all rolled up and tied with a spliced-up rope. When it's time to hit the hay, I simply untie the rope, roll the swag out on the ground, fold back the canvas flaps and climb in like any other bed. Bull ants, snakes, spiders and scorpions are not permitted to walk across canvas. That's just the rule of the bush and for some reason it works.

After a few months exploring the interior and swimming in dozens of ancient waterholes, I landed a job as a mechanic outside of Alice Springs. I wasn't formally trained but in those days riding a motorcycle meant being able to fix it as well. I'd grown pretty good at keeping it running and become quite handy around other machinery too. The workshop at Santa Teresa Mission, 70 km down a dirt track south-east of 'The Alice' on the edge of the Simpson Desert, gave me a chance to see if I was any good. 'How long did it take you to ride out here on that?' was my only interview question before getting the nod to start work.

The mechanics at Santa Teresa were motorcyclists too, with a love of the challenging ride to town through creek crossings, sandy stretches and big, high-speed, corrugated straights. They became mentors who honed my mechanical education by finding out where I lacked knowledge. If I didn't know much about electrics, they gave me every electrical job that came in the workshop until I was proficient.

These men encouraged me to figure things out for myself. On the first day they handed me a box filled with bits of machinery. It was an engine that needed to be assembled. Now, pulling something apart and putting it back together is one thing;

starting from scratch with a pile of random parts is a little more challenging.

The main form of encouragement I received was laughter from the older blokes as I put pieces together only to pull them apart again. I persisted, however, and took great pride in starting that motor. I don't know if that's the sort of thinking they teach at university, but this had been a master's degree in learning how to figure things out.

This dusty, hard-won bush education helped me develop my thinking in relation to astronomy too. Figuring out how to adjust the tracks on a D6 Caterpillar bulldozer is largely the same as understanding how the moon and Sun are geared together to make the tides on Earth. It's just a matter of watching and learning until it all makes sense.

On the weekends I'd down tools and roar out of town to explore central Australia. As ever I'd unfurl my swag and camp under the stars – a mini-observatory for one. Even though I'd started to feel familiar with the glittering roof that stretched above me, the universe would often throw me a curve ball.

One time after work I knew a full moon was due to appear and I was excited about making a night of it. Keen to watch the spectacle from a fitting vantage point, I packed a couple of beers in a backpack, climbed on the Kawasaki and rode out to some tabletop mountains on the edge of the Simpson Desert.

When I reached the top I kicked the side stand down, put my feet up on the handlebars, stretched out on the backrest and cracked open a beer. 'How good is this?' I pondered as I waited for the moon – my ol' pal from my pearl-diving days – to join the party.

I was just in time to watch the Sun disappear over the western horizon. A dazzling display of reds, yellows, mauves and greens followed as I awaited the full moon. Imagine my surprise to see *half* a moon drifting up over the desert mountains. I knew in all my heart that it was to be a full moon! I sat up in surprise, spilling my beer and shaking my head to see if I had missed something, like maybe a whole week.

Just as when I felt the need to figure out how to assemble a box of engine pieces, I was compelled to work out what was happening and why. Right away!

I climbed off the 'Kwaka' and looked behind me to the west where the Sun had just set. Although it had disappeared below the horizon, I knew roughly where it would be at that time. Then I took stock of where I was standing on my own planet and factored that into where the moon was hovering out to the east. I realised the moon would have indeed been full were it not passing through the shadow cast by the Earth.

'It's a lunar eclipse!' I marvelled to myself. The eclipse wasn't a particularly rare event, but to me it was totally unexpected and the lesson it taught me was priceless. 'Pay attention, have a look at what's happening around you, and you'll figure it out.'

My next job ensured I continued my self-tuition from the comfort of my foam and canvas observatory.

By 1985 I was back in the Kimberley, where I scored a job working for a team of 'line cutters' on vast cattle stations along the edge of the Great Sandy Desert outside Derby, WA. Line cutting is the delicate art of driving a D7 bulldozer in a dead straight line for countless kilometres across the wilderness.

We'd gouge massive lines over the sand-dunes and through the creek beds in preparation for fencing. The lines had to be ruler-edge straight. If you veered off and went crooked you had to start again; otherwise the fencing would fall over. It was no mean feat.

I started out as the mechanic but ended up being the dozer operator *and* the truck driver as well. We'd work hard all day, come into the camp just on dark, have a few beers, a feed, and go straight to bed. Lying back under a sky full of stars constantly blew me away.

About this time I developed the custom of waiting for a 'shooting star' or meteor to streak across the blackness before I dropped off when sleeping beneath the stars. I've only ever waited 10 minutes for one and often I'll see three or four. I always sleep the deepest when the roof is a few hundred thousand light years away. By waking before dawn I slowly tuned in to how far the Earth had turned among the stars during the night.

The repetitive nature of operating trucks and dozers put the mechanical part of my brain on autopilot during the day. This allowed the rest of my mind to wander off into the wilds of the cosmos. I'd ponder how and why the solar system functioned the way it did as I rumbled along on the bulldozer, and when night fell I'd peer into the sky to see if I was right. It took a while but each little discovery made me more and more excited.

Driving a road train on the long trip south from Broome to Perth, I had plenty of time to notice the stars climbing higher into the southern sky and disappearing behind me over the northern horizon. On the return journey north, I noticed the movement of the stars reversed. At that moment I realised I was seeing myself driving around the curve of the planet.

Travelling east and west didn't make any difference at all, because the Earth is turning us that way already. Travelling north and south, however, it was very easy to see that my perspective on the sky changed in only a single day's drive.

It became clear that if I wanted to know more, I needed something other than my swag and my own eyes. I got my hands on a copy of the *Little Penguin Dictionary of Astronomy*, an excellent tome that only served to supercharge my hunger for astral knowledge.

The book contained the valuable advice that if you want to learn about the night sky, don't rush out and buy a telescope. It recommended that stargazers first buy a pair of binoculars and learn where things are in the sky before taking the next step. I did what the people at Penguin said and invested in a pair of 10x50 SteinZeiser binoculars.

The benefit of using binoculars when you're starting off is they give you a wider field of vision. Looking through a telescope is akin to observing space through a drinking straw, whereas using binoculars is like peering through a toilet roll. You're able to see a bigger patch of sky.

Binoculars and telescopes work by allowing more light into your eye. A healthy eye will open out to about 6 mm when it's fully dilated, so you're only getting 6 mm worth of light into your eye. With 10x50 binoculars you're putting 50 mm worth of light into that same eye at 10 times the magnification.

By lying back in my swag and gazing up through my trusty binoculars, I was able to see hundreds more stars than I could with my naked eye. Things that once appeared as fuzzy patches of light in the sky turned out to be clusters of stars.

Armed with my binoculars and a simple star chart, I spent

countless nights over the next few years exploring space. With the star chart I was able to learn the location and names of some of the stars I'd been staring at half my life. I figured out that some of them were planets.

I kept up the same sort of work all through my 20s, playing with bulldozers and graders and driving big horsepower Mack and Kenworth trucks pulling triple road trains. I ran mechanical workshops on various cattle stations and in town for other businesses, before starting my own business doing the same. But barely a day went by – and certainly very few nights – when I wasn't wondering about my place in the universe.

Most people – myself included – know intellectually from a certain age that the Earth spins on its axis. Some no doubt have an abstract awareness our planet also takes a year-long journey around the Sun. For most people, however, that's where it ends; with the vague, intangible knowledge that this colossal movement is taking place. Rarely do people consider it as they get lost in the very tangible tangle of day-to-day life.

I completely understand. The fact is you have to spend a lot of time observing the signs of our planetary movement before you truly see it happening for yourself. For one thing, the sky dictates the rate at which you can learn about certain aspects, because it constantly changes during the course of a year.

As my observational skill and experience slowly grew, I realised some of the stars I was getting to know were only around at certain periods of the year. Over time as I learned the sky, I would notice something – let's say Leo – and wonder why I had only just picked it up. Then I realised that I hadn't noticed it before because it simply wasn't there a month or so ago.

I looked at Leo in the sky and then I looked at the Earth that I was standing on, and then I looked at where I knew the Sun to be, and suddenly it all made sense in 3D.

I realised that the Earth had simply come back to the same place on its journey around the Sun to bring Leo back into the same part of the sky, at the same time of year. By putting those three things together – the Sun, Leo and the Earth – I realised the very direction that the Earth was travelling in.

It was at that moment I actually started to feel the Earth moving around the Sun. It was a very real and physical experience and I almost had to rebalance myself on the planet to reassert my bearings.

Not everyone was on the same page with me, of course. Some of my workmates thought I was crackers. Some were *kind of* interested but it took me a while to realise that people weren't as excited about the cosmos as I was. Others didn't even want to know.

I had a mate called Marty. We used to fix Harley-Davidsons, mess about rebuilding engines and going on long rides together. One night after we'd stopped and were setting up our swags, I pointed up and said, 'Hey, Marty, that's Jupiter up there!' Instead of looking where I was pointing, Marty gazed at the ground.

I tried again the following evening. 'Hey, Marty, you seen Jupiter tonight?' He just looked at the ground again. It took me a few days to realise, 'Oh, he doesn't want to know. He's just not into it at all.'

I never bothered to tell Marty where Jupiter was again. It showed me that there are people who want to know and there are people for whom – for whatever reason – the idea of careening through space is too much. I realised that if I wanted to share these things with others, it was important that they wanted to know. Sadly Marty has passed away, but I'm grateful he left me with that gift of knowledge about teaching. It has come in very handy.

Chapter 3

Through the Looking Glass

Saturn, with its shimmering waistband of wafer-thin rings, has fired the imagination of everyone who's had a glow-in-the-dark version of the planet stuck to their bedroom ceiling. Seeing Saturn for real through a telescope is a thrill on a different scale altogether.

The first time I did it, in 1987, I was completely blown away. Of course, I already knew what Saturn looked like. I'd seen drawings, paintings, and photographs of it taken in the early 1980s by the NASA space probes Voyager 1 and Voyager 2. Nothing, though, could have prepared me for the otherworldly sensation of eyeballing spectacular Saturn in person from right here on Earth. What a blast!

Imagine how Galileo must have felt when he became the first one to point a telescope at the night sky over 400 years ago. Although a German-Dutch spectacle maker named Hans Lippershey had invented the telescope in 1608, Galileo Galilei – the brilliant and renowned Italian scientist – tinkered with the idea and built his own version in 1609.

As an astronomer Galileo refined not only his telescopes but his powers of observation. In early 1610 he became the first person to see Jupiter's moons, although he initially described them as 'three fixed stars, totally invisible [to the naked eye] by their smallness'.

He watched Jupiter over the following nights and noticed these 'fixed stars' were, in fact, moving. On the fourth night one of the stars disappeared, which Galileo put down to it being hidden behind Jupiter. A few nights later he discovered a fourth planetoid orbiting Jupiter and he determined that these 'stars' were Jupiter's moons. Today we know them as the Galilean moons – Io, Europa, Ganymede and Callisto.

I still remember the first time I saw them, with Tom Smith of the Perth Observatory. I was amazed at just how far away from Jupiter the moons were.

As he continued to hone his telescopes, Galileo racked up multiple major astronomical firsts, including observing the phases of Venus (similar to our moon's phases) and dark spots on the surface of the Sun, and he was the first person to get a magnified look at the mountainous and cratered surface of the moon.

Although he was also the first person to observe Saturn through a telescope, Galileo's instruments weren't powerful enough to clearly show its rings. Back then, no-one knew the rings even existed, and when Galileo first saw them he mistook them for two large moons, or maybe even ears, on either side of the planet.

When he looked at Saturn again two years later, these 'moons' had bizarrely disappeared from view. Today we know that was because Saturn's rings had gone 'edge on' – an event that happens every 14 years. Galileo was looking at the razor thin rings side-on, so they were invisible to him.

A couple of years passed before he viewed Saturn again, this time finding the 'moons' had returned. With no other explanation and limited by the performance of his telescopes, Galileo concluded the rings were 'arms' of some type. It wasn't until 1655 – 13 years after Galileo's death – that Dutch astronomer Christiaan Huygens correctly described the rings as a disc surrounding Saturn.

I think Galileo would have killed to have one of my telescopes.

My 30s brought on a transition, as it often does for people. I became a father and then a single parent. I settled back in Broome and opened Quicke Spanners, fixing motorbikes and machinery. My daughter slept in a sleeper cab a few times while I was still driving trucks, and when I set up my workshop in town she'd mess about with my spanners as I worked. At six weeks old she spent her first night in a swag under the stars. I love being a parent and consider it a great privilege.

I welcomed another 'baby' of sorts into my life around this time: my first telescope. Scanning the cosmos with binoculars had made me realise that some of the fuzzy patches I'd seen with the naked eye were, in fact, clusters of stars. Over time I found more and more clusters and started to become familiar with their locations in the sky.

There are two main types: open clusters, which are fairly loose groups of maybe 100 or so stars, and globular clusters, which contain hundreds of thousands of stars – sometimes millions. One of the most spectacular globular clusters visible from anywhere on Earth is found in the Milky Way, deep in our southern skies. It's called Omega Centauri and it is home to about 10 million stars. Once I found *that* in the binoculars, I just had to get a closer look.

A mate of mine named Dee discovered that his neighbour had a half-forgotten 4½-inch telescope stuffed under his bed. The guy didn't even know how to use it, so he let us pull it out. Being a mechanic, I quickly figured out how to put it together. I knew where Saturn was directly overhead by now, so we pointed the scope straight at it, just as Galileo had done.

I was gobsmacked. There it was. Saturn is tiny in a 4½-inch telescope, but it was absolutely perfect. I couldn't believe my eyes at the impossible and incredible perfection of Saturn and its beautiful rings.

I knew to look for the shadow of the rings on the planet and the shadow of the planet on the back of the rings and the flattening of Saturn at the poles and the bulging of the equator. I was blown away that I was seeing the real light of Saturn that had travelled for an hour and a quarter across space to arrive in my eye.

Every human should be shown Saturn. It has the ability to change your life. I've seen this happen thousands of times. Standing in Dee's front yard in Derby, I said to myself, 'If this is what I can see in a 4½-inch telescope, I'm going to get a 10-inch telescope!'

I started putting coins in a jar and eventually saved $1400 for a second-hand 10-inch Dobsonian telescope purchased from a dealer in South Australia. The 10-inch was, and still is, a very significant telescope. At first glance it looks more like a cannon suitable for launching small children to the moon. Instead it launched me on a life-long mission of discovery and, unwittingly, on a trajectory into a career as an astronomer.

Although I bought it purely to satisfy my own desire to explore the stars, pretty soon my mates wanted to come around and have a look. I'd take them on mini tours of the astral neighbourhoods I knew. I'd show them the Galilean moons of Jupiter, the rings of Saturn, the surface of the moon and the 10 million suns of Omega Centauri. I explained how Omega Centauri is 200 light years across and 16,000 light years from Earth, which meant the glow they were seeing was emitted about 10,000 years before recorded human history began.

Soon enough, word spread around Broome that Greg Quicke had a big telescope and knew how to use it. One of the local schools asked if I could run some stargazing sessions with the students. 'Sure!' Next thing I knew, the Cable Beach Club Resort invited me to do the same with their guests. 'Yeah, why not?'

I got a real kick out of watching people's jaws drop as they looked into the eyepiece – especially when I showed them stunning Saturn. I had some amazing experiences of my own with the 10-inch too. In July 1994 I took my three-year-old daughter on a 100 km 10-day heritage walk along the remote coastline north of Broome with the local Goolarabooloo people. There were about 100 people on the walk and we were accompanied by a support truck, so I took the telescope along – partly because I wanted to share the joy of stargazing with the local Indigenous people, and partly because I wanted to witness an historic cosmic collision.

During the previous year US astronomers had discovered a comet they named Shoemaker-Levy 9. Its orbit had been bent out of shape by a previous close encounter with Jupiter, and on its next pass it was going to make direct impact. The force of Jupiter's gravity had started to break it apart and it was due to crash into the planet during the middle of that July. It was the first time in recorded history that humankind was going to have the opportunity to witness the extra-terrestrial collision of objects in our solar system. I didn't want to miss that!

I set up the scope each night after walking the ancient coast and took my fellow trekkers on tours of the night sky. I also used the opportunity to closely study Jupiter so I'd be able to notice any changes the comet made to the planet's surface when it eventually crashed.

Jupiter appears as a reasonable-sized disc in a telescope. It has surface detail in the form of cloud bands and a three times the size of Earth red spot – a gigantic storm that rages 22 degrees south of the equator – all of which are easy to see, even with a small scope. With the mighty 10-inch Dobsonian at our disposal, we got a really good look at it.

The collision – or series of collisions, as 21 pieces of the comet rained down – was due to begin in the afternoon of 16 July. About 4 pm on the long-awaited day I stood near to the scope under the blue Kimberley sky and searched for Jupiter with my naked eye. The beauty of Jupiter is you *can* see it during the day – if you know where to look.

Having spent the previous decade consciously watching the movement of the Earth, the planets and the stars, it still took me 20 minutes of peering out into infinity to find Jupiter. It snapped into sight as my eyes focused on the other world that is Jupiter, 884 million km away in an otherwise blue and featureless sky.

Keeping my eyes locked on Jupiter, I walked over to the telescope and manhandled the big machine towards it.

Bang! I got it. There was Jupiter in the field of view of a clear blue sky, with me ready to witness humanity's first cosmic collision.

Over the following hours, 21 pieces of Shoemaker-Levy 9 slammed into the surface of Jupiter at 216,000 km/h. Jupiter spins fast – once every nine hours and 55 minutes – so each piece of the comet hit in a different location: a piece would crash, the planet would rotate a little, and then the next chunk would hit in another spot. As this silent carnage played out, I could clearly see a chain of dark blemishes spread across the surface of Jupiter, which made me an eyewitness to inter-planetary history in the making. I admit I was fairly pumped.

Although the others on the walk weren't familiar enough with Jupiter to be able to see that anything had changed, they picked up on my excitement and listened intently, getting just as charged as I was at such a major event for humanity.

The Kimberley region of WA is blessed with clear skies for around 300 nights a year. This makes it an ideal location for astronomy. Obviously, cloud cover, haze and light pollu-tion get in the way of stargazers the world over. While we're not immune to the effects of light pollution in Broome, it's not exactly Times Square – you can still see the Milky Way if you're standing in the middle of town.

With such optimal conditions on offer, I found myself looking up through the scope most nights. I was also kept reasonably busy conducting casual sessions with schoolkids and the odd tourist group, not to mention my mates, who were always up for a bit of a stroll among the stars.

After a while, my friends started saying things to me like 'When are you gonna get on with it, Quickie?' and 'Mate, what are you waiting for?'

I didn't know what they meant.

'Get on with astronomy, Greg! Do it for a living,' I was told. 'People love hearing what you've got to say.'

Although not 100% convinced of the idea, I took their advice, and in 1995 I started Astro Tours. I put fliers out around town, within two weeks I was booked seven nights a week, and I haven't looked back since. Although I started off with my trusty Dobsonian, I now have a 40-foot shipping container full of big telescopes, ranging from four inches to 20 inches at the Astro Tours outdoor observatory on the outskirts of Broome.

On average 5000 people come to Broome every year and look through these machines while listening to me talk about what I've noticed in all my years looking up at the stars. Although the skies are constantly changing throughout the year, there's always something incredible to behold.

Watching the Earth turn is easy in a couple of hours, and with this knowing comes an ability to be able to watch it do so for the rest of your life. Watching the Earth go around the Sun is possible too, although it takes more like a week to notice it. Setting people up to watch this very real journey also brings me joy.

I feel privileged to be able to share this amazing, fundamental reality with people, and I'm fortunate to be alive at a time when it's okay to do so. When Galileo and other pioneering thinkers started pointing this stuff out 400 or so years ago, they came in for public rebuke for daring to undermine the prevailing view of governments and the Catholic Church that the Earth was at the centre of the universe – full stop. The ruling class didn't want to hear alternative views. When Italian philosopher and astronomer Giordano Bruno proposed in the late 1500s that stars were, in fact, distant suns surrounded by their own planets and that the universe was infinite, the powers that be burned him at the stake.

Were they around today, Bruno, Galileo, Johannes Kepler, Nicolaus Copernicus, Tycho Brahe and all the other great cosmological thinkers of that time would have been delighted to know that not only were they on the right track, they also helped lay the groundwork for even more remarkable break-throughs in human knowledge that were yet to come.

Even up until 100 years ago the generally agreed scientific consensus was that our galaxy, the Milky Way, represented the entire universe. Once again it was an astronomer armed with a telescope who challenged the status quo. But this wasn't just any telescope.

In 1917 engineers completed construction of the 100-inch Hooker Telescope at the Mount Wilson Observatory in California. At the time it was the largest ever built. A few years later the American astronomer Edwin Hubble used the Hooker Telescope to determine that spiral nebulae – a term then used to describe diffuse smudges in the sky – were too distant to exist within the Milky Way. Instead, Hubble reckoned these 'nebulae' were, in fact, entire galaxies outside of our own.

Although he faced opposition from the astronomy establishment, no-one suggested Hubble be tied to a telegraph pole and burned alive in the middle of Los Angeles. Ultimately he was proven correct. It was a major shift in perception for all of humanity and it completely changed the scientific view of the cosmos. One day we thought our galaxy was the entire universe; the next, we found out ours was just one of many in an immensely larger and perhaps endless expanse of space.

In 1990, 37 years after his death, NASA honoured the great astronomer by naming the Hubble Space Telescope after him. This schoolbus-sized telescope was launched into low Earth orbit in order to place it far above light pollution and the smudgy blanket of Earth's atmosphere, and allowed us to see deeper into space than ever before.

Like Galileo and co., I reckon Edwin Hubble would have loved to be around today to see how far we've come thanks to thinkers and scientists like him. In 1995 the eponymous space telescope was used to take a photograph called Hubble Deep Field. NASA deliberately pointed the scope at a very blank piece of sky and took the photo over a 10-day exposure. The tiny section of the sky they imaged – which was about the size of a grain of rice held end-on at arm's length – contained around 1500 galaxies.

A few years later NASA put a different camera on the scope and took another happy snap of space, known as Hubble Ultra Deep Field. This celestial grain of rice yielded 10,000 galaxies. Who knows what the next discovery might be?

Yes, telescopes are pretty awesome, but really I'm just as happy when I'm lying back in my swag, eyes open in wonder as we speed along on our odyssey through space. It's a great first step and will never go out of fashion. I highly recommend it. Tonight is as good a night as any to give it a go for yourself.

Chapter 4

Shrinking the Stars

Like the town of Broome itself, the early days of Astro Tours were pretty loose. I knew I had things to share that people would find interesting, but I didn't have any real structure for just how to do that. In those first days, if I took four or six people out to look through the telescope we could sometimes still be sitting there in the dark at 2 am, discussing stars, space, the Earth and eternity.

Although today when I run a star show, we have a start and a finish time (well before 2 am), much of what I tell people remains the same – simply because the complex choreography of the stars and planets remains the same, like a giant Swiss watch.

During all my years lying under the stars, I asked myself all the big questions: 'How far away are the stars?' 'How big are they?' 'What are they?' 'Where do I fit into the whole picture?' These are the same questions I get from the good people who come to Astro Tours presentations. Fortunately I'd sought the answers in books and journals over the years and had come across some mind-blowing facts and figures to share:

12,756 km

The Earth is more than 12,700 km in diameter. If you drilled a hole straight down from where you're standing now and kept going for 12,756 km, you'd pop out the other side of the planet. That's just another reminder that the Earth is round and that we're surrounded by space on all sides. You'd probably emerge at the bottom of an ocean, given that almost three quarters of Earth's surface is liquid.

3475 km

If you leave our little planet, the nearest place you can visit is the moon. The moon is 3475 km in diameter; about a quarter the size of Earth. This makes the moon about the same width as Australia. The next time you look at the full moon, imagine Perth on one side and Sydney on the other.

400,000 km

The moon varies between 360,000 km and 420,000 km away as it travels around the Earth in its elliptical orbit. Were it any closer, our tides would be very interesting indeed. When I first tried to grasp how far 400,000 km was, I realised that my old 80 series LandCruiser had done 430,000 km. That means that it was officially on its way back from the moon! The nearly 4000 km drive from Perth to Sydney is humungous and takes days, but at least it's comprehensible. Imagine driving across Australia 1000 times!

150 million km

When you move beyond the moon, the distances to other ports of call in our solar system start to get ridiculous, and the numbers end up being almost meaningless: 150 million km to the Sun, between 588 and 968 million km to Jupiter (it varies according to orbit), between 1.2 and 1.7 billion km to Saturn, and a brain-popping 4.28 to 7.5 billion km to Pluto, way out on the edge of our solar system.

I found I needed another way of thinking about these numbers in order to comprehend them, and also to share them with the audiences at Astro Tours. I discovered that shrinking things helped my little earthling brain get a grip on the enormity of it all and allowed me to get it into the heads of my fellow humanity.

1 mm grain of sand

In order to construct a scale model of our solar system, let's shrink Earth down until it's the size of a grain of sand with a diameter of 1 mm. On that scale, the moon is a smaller grain of sand one quarter the size of the model Earth.

31 mm

These two grains of sand are 31 mm apart. Earth and the moon now fit on the palm of your hand, with the moon going around

Earth about once every 28 days. As they're doing this little waltz in your hand, our two grains of sand are also going around the Sun once a year.

109 mm coconut

On this new scale the Sun is a coconut 109 times bigger than the Earth as a grain of sand.

12 m

Instead of being 150 million km away, our coconut-sized Sun is situated 12 m from our grains of sand – about the length of a school bus. (By the way, in this scenario you are seven million times smaller than that grain of sand in your hand.)

Two more grains of sand

Venus and Mercury also travel around the Sun and are closer to it than we are. They are two more grains of sand, with Venus roughly the same size as Earth, and Mercury only one third the size, or not much bigger than our moon. The grain of sand called Venus is 3.5 m closer to the coconut Sun than we are, or about a quarter of the way down the bus aisle. Mercury is another 4 m closer again, getting nearer to the back of the bus than the front.

Since they are closer to the Sun and its gravitational force, these two inner planets get pulled harder by the Sun than we do. That's just the way gravity works. To resist this stronger pull, our sand-sized Mercury and Venus have to orbit faster around the Sun to avoid being dragged into it. The extra speed generates greater centrifugal force, which throws them away from the Sun in elliptical orbits that balance the extra gravity at the closer distance.

The inside lane

Going faster and travelling a shorter distance around the inside lanes (i.e. orbits closer to the Sun than ours) means these inner planets lap the Sun faster than the Earth does. Mercury whizzes around the Sun in just 88 days, catching up with and passing us once every 116 days. Earth's 107,000 km/h speed around the Sun is pretty slow compared to Mercury's 290,000 km/h. Meanwhile, Venus circles the Sun in 7½ months and passes us on the inside lane once every 19¼ months.

The outside lane

The further away from the Sun you go, the slower the planets move through space and the longer they take to complete a single orbit.

Mars is another grain of sand that's half the size of the Earth. It is 6 m away from the sand in the palm of your hand and takes over two years to lap the Sun.

Grapes and peas

Jupiter dwarfs our grains of sand as an 11 mm grape. Earth fits across the face of Jupiter 11 times. Our grape is 60 m away from us and takes 12 years to get around the Sun.

Saturn is a 9 mm wide grape that is nearly twice as far away as Jupiter at 114 m out – put 10 of those school buses end to end. It takes Saturn close to 30 years to do a lap around the Sun.

Uranus and Neptune are both 4 mm peas. Uranus is 230 m away from our grain of sand and takes 84 years to orbit the Sun. Neptune is 370 m out and takes 165 years to do a single lap.

Yes, we are going to keep Pluto. Pluto is nearly half a kilometre away from the palm of your hand and is another grain of sand, but only about half the size of our moon. Pluto is a tiny little world that takes 249 years on its big, looping orbit to circle the Sun. Pluto is so far away from our Sun that there would be no daylight, as we know it, on Pluto. The sky would be completely filled with stars all the time. Our Sun would be the brightest one. Take a moment to imagine that view from Pluto.

Plutoids

During the 1990s a host of Pluto-sized objects that orbit the Sun were discovered. This led to poor little Pluto being downgraded in 2006 from the most distant planet in our solar system to that of a 'dwarf planet' or 'sub planet'. I wouldn't worry too much about this decision, because it hasn't made one bit of difference to Pluto. Pluto is happily doing what Pluto has always done.

At any rate, the largest of the Pluto-like objects beyond Pluto have been classified as Plutoids.

The stars are suns

When you look at the night sky and behold the sparkling pinpricks of light we call 'stars', remember that every one of them is another sun. Every one of those suns is highly likely to have its own family of planets going around it. So every one of those stars could also be called a 'solar system'.

The stars are all suns in the same way that our Sun is a star. The only reason we see our Sun, as we do, as that blazing orb of light in our daytime sky, is because it is really, really close to us – just a trifling 150 million km! That's close enough to warm our skin and even burn it, if we don't shelter from its rays.

To put our model into perspective, if the Sun is a coconut 12 m away and Pluto is a grain of sand half a kilometre away, that would make our solar system about 1 km across.

3200 km

On that scale, the closest solar system to ours is another coconut 3200 km away, with its own peas, grapes and grains of sand orbiting around it. If the Earth was a grain of sand in Broome, and our coconut was 12 m away, about as far as you could throw it, the next coconut with its companions – perhaps also spreading out across 1 km – would be located somewhere across the other side of Australia near Cairns.

That particular coconut is a star called Alpha Centauri. Alpha Centauri is clearly visible in our southern skies as one of the two 'pointer stars' associated with the Southern Cross, which we'll look at later.

In fact, this coconut in Cairns would actually be *two*

coconuts 300 m apart. The solar system we call Alpha Centauri actually has three suns, two of which are our coconuts that take 80 years to go around each other. To us they look like a binary star system, or a double star – a solar system with two suns.

From Earth, the human eye perceives Alpha Centauri as just one star, because these suns are so close together, relatively speaking. It's a bit like driving down a highway at night and seeing an oncoming car appear as one light in the distance. As the car gets closer, the headlights split into two. A telescope does the same to Alpha Centauri. It gets us close enough to split the two stars and we see them as the most beautiful pair of diamonds called Alpha Centauri A and Alpha Centauri B.

The third sun, Proxima Centauri, is slightly closer to us than the two suns we see as Alpha Centauri.

Compared to the two 300 m apart double coconuts of Alpha Centauri, Proxima is a 16 mm grape, 150 km away from the pair. Proxima goes around the two suns of the Alpha Centauri system once every 500,000 years or so. Proxima is a red dwarf star, too dim for us to see from Earth without a telescope. Even from Alpha Centauri A and B, Proxima would be one of the dimmest stars in the sky.

Many solar systems are far more complicated than our own single sun variety with its family of planets. Two thirds of the stars we see in the sky are double stars or triple stars. Some have even crazier combinations.

The model

So now we have a model, which can perhaps allow us to explore certain possibilities and grasp some of the scale of our local solar system and its place amongst its neighbouring solar systems.

Even though Alpha Centauri is the closest star we can see, it is only the third brightest star in the sky. Stars are bright either because they are close, or because they are big. Really big.

Sirius

Sirius is the brightest star in our skies, even though it is twice as far away as Alpha Centauri. In our model it would be a basketball bobbing in the South Pacific Ocean east of New Zealand – about 6400 km away from our grain of sand in Perth. Sirius is a significant presence in our local region of space. It's twice the size of – and four times as bright as – our Sun and dominates all of the other stars for light years around it.

Canopus

If Sirius is a mighty star in our cosmic neighbourhood, then Canopus – shining as the second-brightest sun in our skies – is the grand-daddy-boss of the block. In our model, Canopus is about halfway to the moon and the size of an elephant. Compare that to our grain of sand orbiting our coconut in Broome.

I'll have more to say about Canopus later on. However, for now, a consideration of those grapes, peas and coconuts might just be food for thought on your own journey of understanding of the reality that this model represents.

Chapter 5

Far, Far Away

While it's useful to reduce the Sun to a coconut and Earth to a grain of sand, it doesn't really do justice to the cosmos in all its immense, fantastical, mind-boggling glory. On top of shrinking solar systems and galaxies so they fit inside our heads, we can attempt to comprehend the universe's enormous distances using a different metric; one that has virtually no practical application here on Earth. So without further ado, let there be light speed.

The very notion of a 'light year' has been a source of confusion for many since the term was first used in the 1850s. It sounds like a measurement of time, but it's really a measure of distance. A light year is the distance that light travels in a year. The only difficult thing about a light year is that it is a really, really long way!

Light travels at 300,000 km per second. Therefore a light year is 300,000 km, multiplied by all of the seconds in a year. There are about 31.5 million seconds in a 12-month period. Do you want to do the maths?

Nope? That's okay, I've already done it. To calculate the distance of a light year, you have to multiply 300,000 km by 60 seconds to get a minute, by 60 minutes to get an hour, by 24 hours to get a day and by 365 days to get a year. The total is 9,460,800,000,000 km. To put it into words you're familiar with – but still can't quite fathom – a light year is equal to nine trillion, 460 billion, 800 million km in distance.

I think we'd better just stick with calling them light years.

Alpha Centauri – our nearest star and erstwhile coconuts on the other side of Australia – is 4.3 light years away from Earth. In other words, the light from Alpha Centauri takes 4.3 years to reach us here on Earth across a distance of about 40 trillion km.

By comparison, our nearest neighbour – the moon – is 1.3 light seconds away. With no light of its own, that 'moonlight' started its journey as 'sunlight' that took 8 minutes and 20 seconds to travel from the Sun to the moon, before bouncing off the moon as reflected moonlight and taking another 1.3 seconds to arrive in your eyes.

We see the moon and the planets in the solar system because our Sun shines on them. They don't have any noticeable light of their own. We see the stars because they are suns and blazing with their own light. When Galileo first spied the moons of Jupiter twinkling like 'fixed stars' in 1610, he was in fact seeing the reflection of the light of our Sun shining on them.

Different planets, different schedules

Depending on where they are in their orbits compared to where we are in ours, the distance from Earth to the planets varies considerably, especially with the closer inner planets. Of all our fellow wanderers, Venus comes the closest to Earth at just 2 minutes, 20 light seconds when it passes us on the inside lane. Mars can get as close as 3 light minutes, while Mercury will always be at least 5 light minutes away, even when it's at its nearest point to us.

However, when Venus is on the *other* side of the Sun, it is more like 14.5 light minutes away. When Mercury is on the other side of the Sun, it's 11.5 light minutes away.

The distance between Mars and Earth varies even more, certainly from one side of the Sun to the other, but also whenever it is closest to us at opposition. Opposition is simply when a planet is opposite the Sun in the sky. When it is opposite the Sun, it is on the same side of the Sun as we are. This is as close as we can get to Mars.

Because the Martian orbit is more bent out of shape, or elliptical, than ours, we can get as close as 3 light minutes when opposition happens at Mars' 'perihelion', with Mars at its closest approach to the Sun. This can stretch out to 5½ light minutes if Mars is at the outer bulge of its orbit, or at 'aphelion' – its furthest point from the Sun. When Mars is conjunct the Sun, or on the other side of the Sun from us, the light reflected from the red planet takes more like 20 light minutes to get here. That's a distance nearly halfway to Jupiter!

Jupiter varies between 35 and 51 light minutes from Earth, between opposition and conjunction. Because it's such a big planet, our view of Jupiter is always superb. Saturn's distance from Earth varies from about 70 to 87 light minutes, depending upon which side of the Sun we are on in relation to each other at the time.

If we could travel at the speed of light, it would take less than 20 minutes to get from Earth to all of the inner planets, and to the Sun. A trip to Jupiter would take about 45 minutes, another half-hour to Saturn, and in about 4½ hours we could include Uranus, Neptune and Pluto. We could go to Pluto and back for the weekend!

The catch

In that same imaginary light-speed rocketship, a one-way trip to Alpha Centauri would take us 4 years and 4 months, doing 300,000 km per second. We haven't got a clue how to do that yet. With our current technology, that journey would take us about 60,000 years.

Pondering the size of the cosmos really does flex the mind, and all of us at one time or another have probably tried to comprehend the idea of infinity. Even if you decide there's a brick wall at the end of the universe, pretty soon you start to wonder what's on the other side of the wall.

Even contemplating the distance to the moon is often too much for some people and they need a cup of tea and a lie-down. Ballooning your imagination so it reaches the Sun is another great leap in the mind-expanding experiment that astronomy allows you to experience. The next jump, out to the stars, multiplies this expansion of the imagination massively.

Physics, as we know it, tells us that nothing can go faster than light. I'm going to suggest that there's one thing that can – and that is your mind. It doesn't take you 4 years and 4 months to include the concept of Alpha Centauri in your mind. Well, for some people it might. Others, though, can do it rapidly and travel to these far-flung places. Astronomy is a wonderful tool for mind expansion. Maybe your mind is faster than light!

Stars are here to stay

Whenever I talk to audiences about stars being light years in the distance, someone inevitably asks if it's possible some stars are no longer there, and we're simply seeing the light they emitted thousands or millions of years ago. I think that is something our science teachers told us when they'd run out of real things to say.

Let's just think about it for a minute. Take, for example, the star Achernar – 140 light years away. Or the globular cluster 47 Tucanae – 17,000 light years from Earth. Or the Sculptor Galaxy – 11 million light years away. Could these really have ceased to exist and we're only seeing their remnant light? The story just doesn't stack up. The fact is that 140 years, 17,000 years, 11 million years is nothing in the lifetime of a star, or a cluster of stars, or even an entire galaxy. Scientists reckon even our relatively small Sun is around 4.5 billion years old!

So my answer to this question is always the same. 'The stars you're seeing are still there – you have my personal guarantee.'

They'll be there again tonight too. Why don't you grab your star chart, go look for Alpha Centauri and prove you can see something that's 9½ trillion km away. And that's the closest one!

Chapter 6

Galactic Pizza

One evening in the early days of Astro Tours I was busy setting up my telescopes when a young German fellow casually remarked, 'What a pity it's cloudy tonight.'

I stopped what I was doing and looked up and around. I couldn't see a single wisp of cloud in the crystal-clear Kimberley sky. Bemused and unsure of what to say to the bloke, I stayed silent and returned to tinkering with the telescopes. Then it dawned on me: he was talking about the clouds of stars overhead.

'Mate, they're not clouds,' I informed him. 'What you're seeing is clouds of stars – some of them more than 10,000 light years away. That's the Milky Way, my friend.'

Now it was the German fellow's turn to fall silent – and for his mouth to fall open as he gazed at the lustrous band made by the millions of suns aglow above our heads. Having come from light-polluted Europe, he'd never experienced the Milky Way on the same scale, or with the same clarity, as he did on that dark night in the wilds of the Kimberley.

The Milky Way galaxy – our home galaxy – tends to have that effect on people, and probably has done since the dawn of humanity. The Milky Way takes its name from the Latin term *via* (way) *lactea* (milk); an ancient nod to the hazy white ribbon of light that shines from a dense grouping of stars, most of which you can't even see with the naked eye thanks to the sheer number of them – around 200 billion. While it looks amazing from my Kimberley backyard, the Milky Way galaxy is visible from dark viewing sites all over the world.

From certain vantage points and at certain times, however, it feels like the galaxy can wrap around you in a spine-tingling interstellar embrace. I'll never forget the night I experienced this while working on bulldozers in the Great Sandy Desert back in the 1980s. On a hot, starry October evening, I climbed to the top of a high desert mesa and noticed the country around me was so flat I had to look down from my lofty perch to see the horizon.

I didn't know it then, but at that time of year and at the right time of the night, the Milky Way runs 360 degrees around the horizon. It took a moment or two before I realised that what I was seeing was our own galaxy completely surrounding me. Standing and turning around to take it all in on my lonely desert outcrop, I was once again filled with wonder at my place in the universe and the never-ending opportunities it provides for me to learn something new.

Generally speaking, the southern parts of the Milky Way – from Sagittarius through to the Southern Cross and Carina – are much brighter and richer than the northern stretches from Orion through Cassiopeia. Because the Milky Way extends as far north in our skies as Cassiopeia and as far south as the Southern Cross, it depends on where you are on the planet as to how much of the Milky Way you will see. In the Kimberley we get to enjoy every part of it at different times of the night and at different times of the year.

If you're in the northern hemisphere you simply won't have access to the brighter sections in the southern Milky Way, because they will always be below your horizon. Even the fainter parts of the northern Milky Way that cross your skies will be almost impossible to see unless you can find a light-free and unpolluted site.

It's hard for us to discover too much about our own galaxy because we are embedded deep inside of it. The huge plumes of stars that surround us get in the way and prevent us from seeing many other parts of the galaxy. No wonder humans thought the Milky Way *was* the entire universe until Edwin Hubble burst that bubble over 100 years ago.

We have since figured out that the 200 billion stars that make up the Milky Way galaxy are laid out in a big, flattened disc that is about 100,000 light years across and just 2000 light years thick. By studying other galaxies, we have determined ours is a 'spiral galaxy' and possibly a 'barred spiral galaxy', with a big bar running across the centre and spiral arms curling off the ends of the bar.

When you look at the Milky Way from Earth, you do so side on; like holding a pizza horizontally at eye level from somewhere inside the rim of the pizza. The 'milky' glow is the light from stars beyond stars, beyond stars, beyond stars, until our view is blocked by the billions of stars concentrated in the flat plane of the galaxy.

A gourmet, thin-crust pizza

To fall back on food analogies and scale models, if our galaxy *was* a pizza, it would be a thin-crust gourmet. With a 30 cm base, it would be just 6 mm thick, with a fried egg in the middle. There is a central bulge of stars in the Milky Way that is 10,000 to 15,000 light years thick, which translates to a 4 cm lump in the middle of our pizza.

When you gaze at the band of the Milky Way, you are looking across the surface of the pizza, through the sea of mozzarella with bits of pepperoni and mushroom jutting out. We could see these toppings as star clusters, nebulae and other structures in the entity that is our galaxy.

When you look away from the band of the Milky Way, you're looking at the space above and below the flat disc of the pizza. By the way, Earth might be a tiny speck of anchovy embedded in the mozzarella cheese.

With the galaxy 2000 light years thick, the stars you see outside of the milky band in the sky are only 2000 light years from the stars on the other side of that milky band. They are the stars that are on our piece of the pizza! If you took a galactic bite out of the slice we are on, you would get a mouthful of all of those stars. The stars within the glowing, milky band, however, are more like tens of thousands of light years away.

Although our Sun is everything to us, it's a small-time player on the galactic scale. Our solar system is situated 30,000 light years from the centre of the Milky Way in a relatively quiet neighbourhood between two of the much busier spiral arms of the galaxy. Our home is the Orion spur – a small spiral structure that's 3500 light years across and joining the outer Perseus and the inner Sagittarius arms of the galaxy.

Turning away from the centre of the galaxy, we look out to the Orion spur as that band of Milky Way stretching across the sky through the Orion constellation. Maybe 6500 light years beyond the Orion spur is the Perseus Arm of the galaxy. Looking towards the centre of the galaxy is the Sagittarius Arm superimposed over the Carina Centaurus arm, with the central bulge of the galaxy somewhere behind all of this. The names and places of these galactic arms are the subject of much study. With so much dust and so many stars in our line of sight, we simply can't see clearly where everything fits in the structure of our own galaxy.

However, you can easily see the central bulge of the fried egg at the centre of the galaxy with your own eyes in our southern winter skies. It looks like a big lumpy bit of the Milky Way around the Sagittarius and Scorpio constellations, and is placed perfectly overhead for best viewing during our peak stargazing season in Broome. The weather is also perfect for sleeping under the centre of the galaxy. Have I told you how good it is to lie back in a swag?

The neighbourhood

Thanks to Edwin Hubble, we know we can look beyond our galaxy to others further afield. We can even see them with our own eyes. If you look deep in our southern skies, you can see two distant galaxies without the help of binoculars or a telescope. You will, however, need a dark sky with no moonlight to see what looks like two broken-off pieces of the Milky Way.

These are the Clouds of Magellan – two dwarf galaxies that are only 10% the size of the Milky Way galaxy. While our galaxy is bristling with 200 billion stars, the Clouds of Magellan each have a modest 20 billion stars.

The Large Magellanic Cloud is 170,000 light years away, while the smaller one is more like 200,000 light years away. They are about 75,000 light years apart, and connected to each other and to the Milky Way by invisible streamers of ionised hydrogen. It's also quite likely these two baby galaxies orbit the Milky Way. If ours is a 30 cm pizza, these dwarf galaxies are party pies. I've learned that the only people who know what a party pie is are the Australians. For the rest of you, a scone is about right.

At a couple of hundred thousand light years away, the Magellanic Clouds are close to us as far as galaxies go. The next full-sized galaxy anywhere near us is the Andromeda Galaxy, nearly 3 million light years away. Since its disc is edge-on to us, Andromeda is bright enough to be easily visible under dark skies as a cigar-shaped smudge of light three times longer than the size of a full moon. It is just north of the Great Square of Pegasus. We see it so easily on the moonless nights from September through to Christmas that I usually wait for someone at my Astro Tours show to ask, 'What is that smudge of light over there?' The Andromeda Galaxy is the most distant object humans can see with the naked eye.

If the Milky Way galaxy is a large-size pizza, the Andromeda Galaxy is a giant family size. With nearly 3 million light years between them, our pizzas would be 9 m apart. The two party pies of the Clouds of Magellan would be an arm's length from our pizza, while Andromeda has a couple of party-pie-dwarf-galaxies of its own.

You can see Andromeda's dwarf galaxies quite easily in a telescope, although Andromeda itself is too big to fit in the field of view of most telescopes. Your trusty binoculars are perfect for looking at Andromeda; you'll see it in its entirety and get a glimpse of its companions too.

The Local Group

There are around 50 galaxies in the Local Group which are orbiting a common centre somewhere between the Andromeda Galaxy and the Milky Way galaxy. Most of them are irregular dwarf galaxies like our party-pie neighbours, the Clouds of Magellan. The Milky Way, Andromeda and the Triangulum Galaxy are the only big spiral galaxies among the 50. Sitting face-on to us, the Triangulum spiral has less surface brightness than Andromeda, so it's harder for us to see. Sometimes called the Pinwheel Galaxy, Triangulum is slightly further away than Andromeda and almost in the same direction.

Our Local Group of 50-odd galaxies is about 10 million light years across and travels through space as a group. This group orbits a gravitational centre in space that is also orbited by other local groups of galaxies. This group of local groups of galaxies makes up a 'galaxy cluster' – in our case the Virgo Cluster, which is about 1000 strong and spreads out over some 35 million light years.

Superclusters of galaxies

Galaxy clusters are local centres in superclusters of galaxies. I know, I know – this is all becoming a bit of a blur at this point.

So, to recap, we live on a planet called Earth that orbits a star in the Milky Way galaxy, which is in the Local Group of galaxies, which is in our local galaxy cluster, which is a member of the Virgo supercluster of galaxies, which is a few hundred million light years across.

Got it?

There are other superclusters of galaxies, including the Hydra-Centaurus Supercluster, the Perseus-Pisces Supercluster, the Hercules Supercluster and the Leo Supercluster, to name just a few. But remember, if anyone asks where you're from, you can truthfully tell them you live in the Virgo supercluster of galaxies.

Galactic megaplexes, gigaplexes and soap bubbles

Superclusters of galaxies seem to be arranged in great walls or sheets that almost form a foam, like soap bubbles, of galactic superclusters in space. These walls can stretch one billion light years and are the biggest structures humans have so far identified.

Millions of galaxies, including our own, appear to revolve around what seems to be an empty space 250 million light years away in the direction of the Centaurus constellation.

If our Sun is the powerhouse within our own solar system, and a supermassive black hole performs an even grander purpose at the centre of our galaxy, then the 'Great Attractor', controlling the motion of millions of galaxies, is a force and presence beyond anything we can imagine.

Somehow, from this little planet of ours, we have the gift of being able to observe all of this, the equipment that allows us to study it and the curiosity to contemplate it.

The Great Attractor, too, is quite likely small fry in even greater structures of the universe. The universe is our way of describing all that is. It is everything that we can see and everything we can't see. It is everything out to the observable edge of space where things appear to be moving away from us at the speed of light and everything beyond this edge.

The very fabric of space is unknown. Does it curve around on itself so that you eventually see the back of your own head? Does it go on forever or is there an edge? Is there a brick wall there? If there is, what's on the other side of the brick wall?

Scientists come up with 'answers' to these questions on a regular basis, only to change their minds on a regular basis too. Does it change the universe when they change their minds?

There are so many big questions that as yet we can't answer. I can promise you one thing for sure, though – the Milky Way is an awesome entity to engage with on a crisp Kimberley winter night. It has the power to change your life.

Chapter 7

Flying Mountains in Space

The second year of Astro Tours coincided with the arrival of a delightful and unexpected visitor. Comet Hyakutake had been discovered on 31 January 1996, by an amateur Japanese astronomer named Yuji Hyakutake. An intrepid stargazer, Yuji had been scanning the skies with a thumping big pair of 15 cm binoculars when he stumbled across it. As astronomers around the world studied it, it quickly became apparent the 4.2 km wide object would come closer to Earth than most other comets had in 200 years. As it raced towards us, Comet Hyakutake put on such a spectacular show it became known as 'The Great Comet of 1996'.

The first I got wind of it was in a phone call from a senior astronomer named Peter Birch at the Perth Observatory around 2300 km south of my bush observatory in Broome. Peter had a favour to ask.

'Greg, there's a comet in the sky but we've got clouds down here,' he began. 'We want you to get a picture of it for us if you can.'

As usual there were no clouds in beautiful Broome at that time of year. 'Yeah, sure, Peter,' I readily agreed, 'but how do I do that?'

Peter talked me through it. 'Get your SLR camera, put 400 film in it, stick it on your tracking telescope and open your shutter for 10 minutes. But first, you'll need to go and find the comet.'

With Peter's help I soon located Comet Hyakutake. It appeared high in the eastern sky as a little fuzz ball, barely visible in my binoculars, but when I checked it the following night it had grown considerably. Over the next few nights it sprouted a tail and became brighter. By early March it was clearly visible to the naked eye. I snapped away as instructed and over ten nights I captured a series of photographs for the team at Perth Observatory.

The change in Hyakutake as it screamed across space was breathtaking. On 25 March 1996, it flew closer to Earth than all but a handful of comets in recorded history. Hyakutake's head appeared as a bright ball on the northern horizon while its blue-green tinted tail extended 90 degrees right to the top of the sky.

There just so happened to be a tourism and industry conference in full swing at the Cable Beach Club at the time. Being relatively new on the tourism scene, I figured I should attend. I also figured I should let the good people of Broome know about the spectacle that was unfolding in our celestial neighbourhood. When I got an opportunity to address the conference, I grabbed a microphone and spoke.

'Right now there's an enormous comet present in the sky and it's passing very close to Earth,' I said during an open session.

'Anyone who's interested should come out and see it with me tonight. You will not regret it.'

There was a fair bit of laughter and people got stuck into me. 'Turn it up, Greg!' they jeered. 'You just want to promote your business!'

Well, it *was* a business conference. But that night I took a busload of people from the gathering to nearby Gantheaume Point – a darkened peninsula jutting out into the Indian Ocean that's pretty good for stargazing. Having already downed a beer or six, some of the guests were a bit rowdy, but when they spilled out of the bus with their eskies, every single one of them fell dead silent as Comet Hyakutake arced halfway across the sparkling sky.

Good for business it was, and even better was my joy in seeing all of those people snapped into a place that was well beyond their stubby of beer.

I, for one, could not get enough of the Great Comet of 1996. When I arrived home later that night, I urged my housemate Jason, who'd just bought a Harley-Davidson (I owned one too), to join me on a ride under the stars.

We roared out of town on our Harleys and barrelled down the highway for about 40 km. When we reached a spot that felt like the middle of nowhere, we stopped in the centre of the road, got off our bikes and lay down in the inky blackness of the desert. The bitumen against our backs was still warm from the rays of our nearest star, the Earth was turning us through space at 1333 km/h and the comet was rocketing headlong through our little slice of the universe. What a trip!

As our eyes adjusted to the dark, we started to see amazing detail appear in the comet's tail. We saw the structures within the tail twist and writhe like a braided hairpiece wafting on the breeze. I could hardly tear my eyes away from it.

A few days later the Comet Hyakutake was gone. The visit was so fleeting that not many people got a good look at it, particularly in the southern hemisphere. Many astronomers – like my mate Peter Birch in Perth – were thwarted by cloud. Although the comet was more visible in the northern hemisphere, we got an excellent view in Broome since we have such good access to northern hemisphere skies at only 18 degrees south of the equator.

Although it was comparatively small, Comet Hyakutake came within 0.1 AU of Earth, a mere 15 million km or one 10th of the Earth–Sun distance. (An Astronomical Unit is the average Earth–Sun distance of about 150 million km.)

Comets are flying mountain-sized objects travelling in highly elongated elliptical orbits that bring them to the inner solar system for close encounters with our Sun, before rocketing back out to the far reaches of our Sun's influence and sometimes beyond.

Like the planets, comets are all doing their own thing. Some are periodic, meaning they return again and again. Some are one-time visitors to the solar system as they blast in and then out again, never to return. Others come for a visit and decide to join the solar system by staying in orbit around the Sun. Halley's Comet goes out as far as the orbit of Neptune and back to the Sun in a 76-year cycle. Others venture well beyond Pluto, taking thousands of years before they return to our skies.

Initial calculations had Comet Hyakutake returning in about 17,000 years from now. In going around the Sun, however, its orbit became a little bent out of shape, or 'perturbed', and it will be more like 100,000 years before we see it again. For this reason, it is considered a 'long-period comet'.

Comets spend most of their time in the outer reaches of the solar system where they are inactive and invisible to us, possibly in a theoretical place known as the Oort Cloud stretching out to maybe a whole light year.

As one of these flying mountain-sized chunks of what looks like rock heads in towards the Sun in its highly elongated elliptical orbit, it reaches a point where its interactions with the Sun's environment cause it to become active. This activity shows up as a huge coma, or cloud of material, surrounding the nucleus.

With a typical comet nucleus being 10 km across, the coma of dust and rocks and gravel and energised glowing, charged, ionised material streaming off the surface of the comet can be planet-sized in comparison. Interactions with the stream of charged particles that we call the solar wind, streaming off from the Sun in every direction, push a comet's tail in a direction away from the Sun, defying whatever direction the comet is actually travelling in. The coma and tail of an active comet in the inner solar system can easily be the biggest object in the entire solar system, including the Sun itself.

The solar wind's effect on the tails of comets gives us a glimpse of the very real substance of the Sun.

My mate Jason and I saw this effect with our own eyes when Comet Hyakutake passed over as we lay on the desert highway in 1996. The twists and turns we observed in its tail were evidence of manipulation by the solar wind. But because the comet was so close to Earth at the time, it's possible our planet's magnetic field played some role in the beautiful display too.

Hyakutake wasn't the first comet I'd seen. Ten years earlier I got a glimpse of perhaps the most famous frequent visitor of all – Halley's Comet.

Measuring 15 km by 8 km, Halley's Comet is visible in the night sky every 76 years and has been observed and recorded by astronomers since at least 240 BC. It's likely, though, that it has been blowing minds since the dawn of humanity.

Its appearance in 1986, however, was a great disappointment to many. There had been a huge lead-up to its arrival; media and others trying to cash in on the event had beat the whole episode up into a frenzy.

The hype had been well deserved during Halley's prior visit in 1910, when the comet sailed very close to Earth and put on a memorable display. The Earth actually passed through the comet's tail on that occasion, sparking mass hysteria, thanks to reports that the entire planet would be poisoned. Instead, this great comet filled the night sky and was a source of wonder for millions.

When it returned 76 years later, Halley's passed the Earth at a much greater distance and thus appeared so small that many people had a hard time finding it. The comet and Earth were on opposite sides of the Sun in February 1986, which led to the least-favourable viewing conditions of Halley's Comet in around 2000 years.

The closest Halley's came in 1986 was 0.42 AU, or about 63 million km. With an increase in light pollution from the urban development, millions failed to even see it. Luckily I was driving a big Kenworth truck carting bulldozers around the Great Sandy Desert at the time, far from towns and the nuisance glow of suburbia.

As ever, I slept underneath the stars every night as Halley's approached. I had no access to the media nor any up-to-date information as to where to find it in the sky. I knew the comet was due to appear, but as I stretched out in my swag I had no idea where to look or what I should expect to see.

On a late-night drive between the remote Sandfire Roadhouse and Broome, my colleague Mick Watt and I stopped to stretch our legs. The sky was a magnificent black and diamond studded affair. Mick and I happened to look up at the same time, and we went 'Oooohhhhhh' at the same time too! There it was – the famous Halley's Comet, very obvious against a moonless, star-filled Kimberley sky. It was my first look at a comet and one I'll always remember.

Having searched in vain for it over the preceding months it was very exciting to finally see Halley's, even if it really only appeared as a small fuzz ball with a tail. The comet was twice as far away as the Sun, and in its least favourable appearance for hundreds of years, but just knowing I was looking at a visitor from the outer reaches of the solar system was enough for me.

About a year after Comet Hyakutake thrilled us with its shimmering tail and near proximity to Earth, an even bigger deep-space traveller decided to swing past our little planet.

At somewhere between 40 km and 80 km wide, Comet Hale-Bopp dwarfed the diminutive Hyakutake and the mid-sized 'celebrity' Halley's Comet. Discovered in mid-1995 by two American observers – Alan Hale and Thomas Bopp – it was due to enter our inner solar system in mid-1997.

Hale-Bopp was by far the biggest comet seen in modern times. While the northern hemisphere got the best of it and southern Australia glimpsed it only briefly before it headed over the northern horizon, we once again got a terrific view from Broome.

Because it passed Earth at a distance 20 times that of Hyakutake (or twice as far as the Sun), Hale-Bopp hung around for longer and appeared nightly over Cable Beach for about six weeks to put on a beautiful display, with its tail stretching nearly 10 degrees across the sky. Watching it change shape, size and position gave us enormous insight into this mightiest of visitors.

I set up mounted binoculars and telescopes on the grass above the beach and invited the general public to take a look. I took the opportunity to hand them an Astro Tours flyer as well.

As with Hyakutake, you didn't need anything other than your eyes to see Hale-Bopp. I was able to share some of my understanding about the comet with hundreds of people who'd come to Broome to experience the famous sunset at Cable Beach. Hale-Bopp was a bonus I'm sure they remember just as well as I do.

I wonder if anyone will line up on Cable Beach when Hale-Bopp – the Great Comet of 1997 – returns in the year 4385.

Who knows – the next flying mountain destined to light up our skies might be discovered tomorrow.

Chapter 8

Rocky Four

I'm sure you've seen pictures of the plane of the solar system in textbooks at school. It is generally drawn as horizontal concentric circles around the Sun, indicating the orbits of the planets. This age-old, disc-like model of the solar system is usually drawn flat on the page, like a record on a record player, with the Sun in the middle and the planets arrayed around it on their individual orbital grooves.

While the planets do indeed orbit the Sun in the same direction and on roughly the same plane, the vinyl record of the ecliptic isn't lying horizontal – it's standing up!

How far it is standing up will depend on your location on Earth. For us here in Broome it spends quite a bit of time vertical or running across the highest part of the sky. The further you go from the tropics the more it will be lying down. However, it will never be lying flat around the horizon unless you're on the Arctic or the Antarctic Circle on the days of the solstices. In other words, hardly ever!

So why do they draw it like that? I don't know, although I can only imagine that maybe it's because the artist never stepped outside to take a look. From our point of view on Earth, all of the planets, including ourselves, travel around the Sun along a very particular path in the sky called the ecliptic. The word ecliptic simply means 'the plane of the solar system'.

We all travel in the same plane and go in the same direction around the Sun. The ecliptic is the only part of the sky that you will ever find planets, or the Sun or the moon. They will always be lined up with each other across the sky. That makes the planets very easy to find, especially if you can picture where the Sun is, and where the moon is.

Simply draw a line through the Sun and the moon, extend it across the sky and below your feet and the planets will be somewhere along that line, even if it is below your feet. If they are above the horizon, the five visible planets are usually bright enough for you to be asking yourself, 'I wonder if that is a planet?' Very often the answer will be yes. Once you find a planet and confirm its place on the ecliptic, where you can trace a line across the sky to include the Sun and the moon, you will be onto it.

So the ecliptic, the only place you will ever find the planets, is an important enough part of the sky to have its own set of constellations marking it out. You know them already. They are the constellations of the zodiac! So if we talk about the ecliptic, or the plane of the solar system or the zodiac, we are talking about the same part of the sky.

Once you know the ecliptic and start to learn the zodiac constellations as a way of finding it, you'll wonder why someone didn't show it to you a long time ago. Lots of people who come to my astronomy shows tell me exactly that. Maybe they did show you – using a school textbook that drew it flat on the page, so you had no chance of ever finding it.

Another question people ask me about the ecliptic is, 'Why are the planets all lined up along a plane like that, anyway?'

My answer is, 'They just are.'

We have the theories of accretion and theories of the preservation of angular momentum which seek to answer these questions and to keep astrophysicists in a job. There are competing theories too, with a big part of our education system geared towards the debating of the state of reality. I guess the religionists and the scientists exemplify this spirit of unco-operation.

I'm very much a wait and see type of person. I'm happy to see the obvious things in the sky and to observe them until they give up their secrets. I have the patience too for my understanding to take lifetimes to achieve. Hopefully, by then we'll have upgraded to the correct orientation for diagrams of the ecliptic plane as well.

Not everything in our solar system orbits the Sun on the ecliptic. There are plenty of rogues out there; asteroids and comets that can swing in from any direction and angle you care to imagine. The planets, though, they're all geared to turn around the Sun on the solar system's great spindle.

Although they all share a relatively close and common trajectory as they tumble through our local area of space, the planets that make up the solar system are very different to each other.

The four small planets out to Mars, including the Earth, are terrestrial or rocky worlds. They move quickly around the Sun and quite noticeably through the sky each night. You'll see their movement as they pass the fixed stars and the other planets easily, as soon as you make the decision to start watching them. All it takes is a little stretching of your mind to include them in your awareness.

Mercury

A fast-moving little planet not much bigger than our moon, Mercury zips between Earth and the Sun every three months or so, lapping us on the inside lane of the ecliptic. Along the way it appears to pass every other planet in the sky, which is why the ancient Romans hailed Mercury, otherwise known as Quicksilver, as the 'Messenger of the Gods', mythically blessed with winged feet and good looks.

Mercury appears in the sky for only a few weeks at a time, while he's either coming at us from the other side of the Sun in the evening western sky, or blasting away from us towards the other side of the Sun in the morning eastern sky.

If you live south or north of the tropics, you'll only glimpse him for a few days at a time. Closer to the equator the ecliptic is generally higher in the sky, so Mercury appears higher above the horizon, meaning those few days stretch out to weeks. The fact is it's rare to see this little planet at all and very few people can say they have.

Mercury isn't as bright as Venus, although he does stand out enough to be seen as a bright star-like object. You always have to look in the direction of the Sun to see Mercury, because he's always close to it from our viewpoint on Earth. However, he's only visible for 20 or 30 minutes in the evenings as he follows the Sun below the horizon.

The same applies in the mornings when Mercury rises

immediately before the Sun and then quickly fades out of view, as the Sun's enormous glare swamps all other light for the day.

Like all four of the inner planets, Mercury is a rocky sphere. Images captured by the spacecraft Mariner 10 in 1974 and 1975 – and more recently by Messenger in 2008 – revealed a heavily cratered planet with a surface similar to that of our moon. Since it is so close to the Sun, almost all of Mercury's atmosphere (if it ever had much of one) has long been stripped off by the intensity of the solar wind. No atmosphere means no wind, no rain and no erosion to blunt Mercury's sharp peaks and jagged mountain bluffs.

Mercury's orbit around the Sun and its rotation around its own axis have an amazingly synchronistic and harmonious relationship with each other. The little planet spins only three times as it goes around the Sun twice. It takes Mercury 88 Earth days to orbit the Sun once, and in that time, it spins just 1½ times. After another 88 days, or one lap later, Mercury has spun on its axis exactly three times.

Were it possible, life on Mercury would be very interesting. If your house was on the sunny side of the planet at midday, 88 days later it would be midnight and Mercury would be back to where it started in relation to the stars. It would have done one lap of the Sun, or completed one Mercurian year.

After another Mercurian year had passed, or in another 88 days, it would be midday at your house again and Mercury would be back to its celestial starting place.

If you stepped out of your house to watch the sunrise from the front veranda, 88 days later you'd be watching the sunset from the back veranda and you'd have to wait another 88 days for the next sunrise.

I reckon building codes on Mercury would demand your house had some heavy-duty insulation, though. During its 88 days of 'daytime', Mercury's surface temperature soars up to 420°C, or 7.5 times hotter than the hottest day ever recorded on Earth. If you wanted to bake scones or party pies, you could simply stick them outside the door for a few minutes and watch them brown beautifully.

If you did manage to judge the timing just right and cook the scones to perfection, they could go back out onto the veranda and into the deep freezer during Mercury's 88 days of 'night-time', when the surface temperature drops to minus 180°C – twice as cold as Antarctica's lowest-ever recorded temperature.

And yes, I have contemplated what life would be like if we were able to survive on Mercury. There's no surf there, so for kicks I would take up dirt-bike riding. However, with no air or oxygen to mix with the petrol in the combustion engine, an electric bike would be necessary. After all, there's plenty of sunshine for 88 days at a time so solar panels on my helmet would provide plenty of juice.

With 88 days of night to take on, I'd probably abandon my house and live out of a swag on the back of my bike. I could ride around on the sunny side of Mercury or, even better, hang around the 'terminator' – the twilight line between night and day – as a way of regulating my temperature and staying comfortable. The terminator moves very slowly, making it possible to spend all day dirt biking and exploring the cratered and mountainous terrain, before ducking over into the night-time side of Mercury for some shut-eye.

Perihelion and aphelion

Mercury's orbit is more eccentric than any other planet, with the exception of Pluto. It is 46 million km from the Sun at perihelion (its closest point) and nearly 70 million km from the Sun at aphelion (its furthest point). Because gravity varies directly in relation to the distance squared, there is more than twice as much gravity from the Sun acting on Mercury at perihelion.

The two laps and three spins resonating orbit and rotation of Mercury slams Mercury right into a gravitational well every time it reaches perihelion. Two points on opposite sides of Mercury alternate in pointing directly at the Sun at each perihelion. It's almost as if Mercury clicks through this gravitational deep point every orbit in the same way a magnet clicks onto iron. Or maybe it works something like the way a racing magneto induces a high voltage in its copper windings, when a spinning iron magnet flicks past them to send a fat blue spark to ignite a heady mixture of racing methanol. My speedway bike-racing days are inspired by Mercury!

It's lucky for Mercury, and for the rest of us in the solar system, that it is travelling faster at perihelion than at any other point in its orbit – about 212,000 km/h. At this speed it blasts through this gravitational well of perihelion and gains enough boost to fling it all the way out as far as aphelion, where it has slowed to just under 140,000 km/h. The Sun's gravity then pulls it back in, speeding it up as it goes. Our racing magneto also meets its biggest resistance at this flux point and is also spinning at its fastest right there to carry it through and make a spark with the energy that would have otherwise stopped it spinning.

Mercury gets a little 'bogged' each time it drops into the perihelion gravity hole and the perihelion point itself gradually moves around the Sun in relation to the stars.

Mercury's resonance with the Sun and key role in the solar system is one we'll see repeated by all of the planets. They each have a unique character that is vital to the functioning of the whole.

Venus

The next time you gaze at the Evening Star, try to remind yourself you're looking at a planet that's almost exactly the same size as the one you're standing on. If you were on Venus looking back at Earth, it would look just like Venus does from here, the only difference being that Venus appears completely white to earthlings. Were you able to see Earth from Venus, you'd likely see the blue of her oceans, the green and brown of her continents and the whiteness of her clouds.

Venus's dazzling alabaster sheen is due to the fact she is always completely smothered in white clouds made up of mostly carbon dioxide, with a thick layer of sulphuric acid crystals on top. This highly reflective layer makes Venus brighter than anything else in the sky, aside from the moon and the Sun.

While the Earth takes 12 months to go around the Sun, Venus completes her journey in only 7½ months. After all, she, too, is on the inside lane from us, so she has a shorter route. Venus is also moving faster than we are, about 125,000 km/h. Our combined motions means that Venus overtakes us on the inside lane on a regular basis, passing between us and the Sun once every 19¼ months.

We see her in the west as the magnificently bright Evening Star for eight months at a time. During this period Venus is coming towards us from the other side of the Sun, growing bigger and brighter the closer she gets. After eight months charging in our direction, Venus seems to vanish as she drifts between us and the Sun. She essentially quits being the Evening Star and it will be 11 months before we see her back in that mode again.

Inferior conjunction

Although she disappears from our evening sky, we only lose sight of Venus for less than a week as she very quickly passes between Earth and the Sun. Then, if you pay attention, you'll see her reappear in the eastern dawn sky, rebadged as the 'Morning Star'. This alignment – when you can draw a line through Earth–Venus–Sun – is called inferior conjunction.

As the Morning Star, Venus races away from us towards the other side of the Sun and leaves us behind in our slower outside lane of the ecliptic racetrack. For eight months she appears as the dazzlingly bright Morning Star just before dawn in the eastern sky. Eventually, she will disappear out of the morning sky as she makes it all the way over to the other side of the Sun from Earth.

Superior conjunction

When she gets there, we have an alignment of Earth–Sun–Venus, with Earth and Venus being on opposite sides of the Sun. With both planets going as fast as they can around

71

the Sun, the race around the ecliptic takes an interesting twist.

Even though Venus is the faster planet, zooming around the inside lane, it takes her another three months to make up enough ground on the Earth to once again reappear in our early evening western sky as the Evening Star. Think of it like two kids racing each other around a tree, with Venus, only very gradually, winning that race.

Venus hangs around the evening sky just long enough for casual observers to expect her to be there all the time. She starts her eight-month reign as the Evening Star low and small in the western evening sky. As the months roll by, she gets higher and bigger and brighter until, just before inferior conjunction, she is so unbelievably bright in the sky that my phone rings off the hook with UFO reports.

Of course, this evening-morning 'star' is a wandering planet and although it is one of our nearest neighbours, Venus's milky-looking cloak has made it hard to study through telescopes.

Since the early 1960s, a number of Soviet Venera and US Mariner space probes have peppered Venus with exploration missions, with various degrees of success. Because thick cloud cover prevents us from seeing the ground, most of our knowledge of the surface of Venus is from radar imaging carried out by the Magellan space probe in the early 1990s.

Magellan unmasked glittering Venus as an incredibly hostile volcanic environment, with an atmospheric pressure equivalent to being 1 km under the sea on Earth. With a surface temperature of around 460°C, she is twice as hot as your kitchen oven on full blast and hotter than the surface of Mercury's baking daytime side.

A day on Venus is longer than its year. Although it takes 7½ months (or 224.7 Earth days) to go around the Sun, Venus takes just over eight months (243 Earth days) to spin once on its axis. Oh yeah, and Venus is the only planet that spins backwards compared to all of the others.

This unusual behaviour becomes even more bizarre when you look at how Venus interacts with the Earth as it makes its journey around the Sun. Every time Venus passes us on the inside lane, she turns her same face in our direction. She does this passing manoeuvre once every 19¼ months, or 584 Earth days.

To add even more to this synchronicity, the overtaking points when Venus lines up between Earth and the Sun are at five specific points in relation to the stars. These five points trace out a perfect five-pointed star, with each point falling in particular signs of the zodiac as seen from Earth (more on this later).

If they meet up at inferior conjunction in Aries, then 19½ months later the Earth and Venus meet up at the next inferior conjunction 210 degrees further around the zodiac in Scorpio. Another 19½ months later and another 210 degrees, they meet up in Gemini, then Capricorn and then Leo, before starting the next formation in Aries again.

Incidentally, any mechanic will tell you this is the same pattern you should use to tighten the five wheel nuts on your car, crossing from one side of the circle to the other until all five nuts are tightened evenly.

As Earth travels 570 degrees around the Sun, Venus travels 930 degrees. In other words, Earth does a bit more than a lap and a half of the Sun, while Venus does a bit more than 2½ laps in order to catch up with and pass Earth each time at one of the points on our five-pointed 'star'.

Venus takes eight of our years and 13 of its own years to pass inside the Earth five times and complete this 'star'. While all of this is going on, Venus spins on its own axis just 12 times.

Each time Venus passes the Earth, on the five points of our 'star', we are both on the same side of the Sun. We are naturally at our closest approach to each other to such a degree that there is a physical and energetic bridge between our two planets. This bridge spans some 40 million km.

The bridge is made of the material stripped off Venus by the ravages of the solar wind. Venus has very little in the way of a magnetic field so the solar wind easily blasts a constant stream of charged particles, or ions, from the top of Venus's atmosphere and pushes them in a direction away from the Sun. When Venus and Earth are aligned with the Sun, we fall right into that stream of charged particles which constitutes the very substance of Venus.

We pass through this bridge and have direct physical and energetic contact with Venus on each of the five points of the 'star'. See if you can feel it when we arrive at the next star point.

If I lived on Venus, I'd want some kind of a winged jet-ski that could fly or swim or paddle through the thick atmosphere to get me up into the cloud tops, so I could see the stars. I'd also like a holiday house that floated on top of the sulphuric-acid clouds because hanging around on the ground would be gloomy and hot, and the ground would quite likely be a bit gooey and semi-molten.

Mars

'That's Mars, isn't it?' is something I hear most nights on my star shows. Most of the time I get to answer 'No'. Every red star in the night sky has been accused of being Mars at one time or other. Indeed, Antares – the red giant star in the constellation Scorpio – even bears a name 'anti-Ares' or 'rival to Mars'. The fact is, though, Mars is often so dim that you'd be hard pressed to notice him at all, as he's one of the most elusive of the five visible planets – Mercury, Venus, Mars, Jupiter and Saturn.

Because Mars circles the Sun once every two years, Earth completes one lap of the Sun as Mars goes halfway around. We have to go around for another year before we catch up with and pass Mars again on the inside lane.

Mars spends quite a bit of time on the other side of the Sun from us where we don't see him at all for around five months at a time. It's similar to the dance Earth does with Venus, only in this case it's Earth that has to catch up with the slower-moving Mars.

When we do start to catch up, Mars appears first in our early morning sky as a very dull little red dot way out on the other side of the Sun. We know Mars is on the other side of the Sun at this time simply because we have to look in the same direction as the Sun to see him.

It takes a long time for Earth to make up ground on Mars. For this reason Mars spends many months low down in our morning sky, a long way away and really quite uninspiring to look at – even through a telescope. It's a whole nine months until Mars appears over the eastern horizon before midnight and we start to see a little action and closer approach from this most secretive of all the planets.

Mars Opposition

We see Mars before midnight for just two months when he appears to rise earlier and earlier each night as Earth reels him in for an overtaking manoeuvre called Mars Opposition.

'Opposition' refers to the period when Mars is opposite the Sun in our skies. This is our closest approach to each other, and for about six weeks, centred on opposition, Mars is spectacular and worthy of the cry, 'That's Mars, isn't it?' With us both on the same side of the Sun and the closest to each other in our 26-month cycle of interaction, Mars shines brighter than any star or planet except Venus.

During this time Mars rises at sunset and you can easily see the line-up of Mars–Earth–Sun for yourself. Mars stays in the sky and shines brightly all night. This is the only time that we get to see Mars up close and with any detail at all.

In a telescope the polar ice caps start to show themselves, and as Mars turns on its own axis, dark areas that you could imagine to be continents rotate into view.

Two months after opposition, Mars sets before midnight and starts to look quite dull. In fact, he's hard to find unless you know exactly where to look. Mars stays in the evening sky for around 10 months after opposition. As the faster-moving Earth races away from him, he gets dimmer and dimmer until we eventually leave him to wallow on the other side of the Sun, where we lose sight of him entirely for another five months.

So our observing cycle in relation to Mars is: five months of missing in action; nine months of early morning, eastern sky, barely visible appearances; two months of civilised hours observing before midnight; six weeks of glorious opposition viewing; and 10 months of gradually being left behind in our evening skies. Mars keeps his secrets well and teases us by staying out of range most of the time. Even when he is at opposition and close by, Mars often throws up dust storms, making telescopic observation of his dusty red surface difficult. People often ask me why Mars is red. I tell them it's because he's got the same coloured dirt as Broome!

Finding Phobos and Deimos, the two moons of Mars, took until 1877, when Asaph Hall, with encouragement from his wife, Angelina, found them using the US Naval Observatory's 26-inch refracting telescope still in use today. At 22 km across, Phobos whips around Mars three times a day at only 6000 km above the ground. Orbiting faster than the daily spin of Mars, Phobos rises in the west (instead of the usual east) and takes 4 hours, 15 minutes to cross the sky.

Doing this nearly twice a day, Phobos is so close that it would be noticeably bigger when it is overhead compared to when it is on either horizon. It would appear no bigger than one third the size that our moon appears from Earth.

The smaller 12 km wide Deimos is more like 20,000 km high and takes 30 hours for an orbit. This is only a bit slower than the 24.6 hour daily spin of Mars, so Deimos takes nearly 5½ days to cross the sky. It is so small and far away from Mars that it would appear to be almost starlike – about how Venus appears to us from Earth.

Being relatively close to Earth (256 million km at the nearest approach), Mars was an obvious target for early space missions by both the US and the Soviet Union. NASA's Mariner 4 did the first successful flyby in 1965, returning our first pictures of Mars as a rocky, lightly cratered, dry world with a thin atmosphere. Even though we can't see Phobos and Deimos from Earth with any ease, NASA's Mariner 9 had a good look at these two tiny pockmarked moons in 1971. The Soviets had managed to put two probes on the Martian surface in 1971, although both failed a few seconds after landing.

In 1976, however, NASA touched down two Viking landers that gave us our first close-up views of an arid, dusty and cold desert landscape. Since then, a large number of missions have mapped Mars extensively from orbit and examined rock samples and photographed some amazing terrain, using a number of highly innovative robotic rovers and some truly amazing landing techniques.

Bursting dramatically out of the Martian landscape is Olympus Mons, the biggest volcano in the solar system. At 27 km high and 550 km wide, it is three times higher than Mount Everest. The Valles Marineris is an incredible 'scar' that runs a quarter of the way across the face of Mars. It's the biggest crevice in the solar system at 4000 km in length, 200 km wide and up to 7 km deep. It would cut Australia in two lengthways and do the same to the USA.

There's plenty of water on Mars, with evidence of surface water flowing in previous ages, although most of it now seems to be locked up in the extensive polar ice caps.

The presence of water would make it easier for us to eventually colonise this most Earth-like of the planets. There is wide interest around the world, with NASA and the European Space Agency both planning crewed missions to Mars, and an active Mars Society seeking to do the same.

Water provides for life and also for fuel, when broken down into its hydrogen and oxygen components. Knowing that there is water present on Mars gives us hope of not only finding life, but also sustaining human life. With water and an atmosphere of mainly carbon dioxide, we could conceivably grow food on Mars that would in turn produce the oxygen that we require to start a colony.

The Martian day is very similar in length to our own and its tilt is tipped over slightly more than ours, so daily and seasonal conditions on Mars would be at least a little familiar to us. Having Mars around the other side of the Sun for months at a time would pose some serious communication problems with Earth, although the very nature of the mission would mean total independence with no hope of any quick rescue from Earth.

And of course there are no liquid oceans on Mars, so no chance of a surf. I'd be packing the dirt bike again.

Chapter 9

It's a Gas, Gas, Gas, Gas

The next four planets out from the Sun are known as the 'gas giants'. These behemoths are essentially planet-sized atmospheres completely devoid of a solid surface. Mighty Jupiter is the closest one to us. He is the most enormous structure in the solar system outside of the Sun.

Jupiter

A giant among the gas giants, Jupiter's volume is a whopping 1300 times that of Earth. If you could gather all of the other planets of our solar system together and stuff them inside Jupiter, it would still have room left over. Unsurprisingly Jupiter is big, bright, brash and super reliable, as he rolls endlessly around the Sun.

We only lose sight of Jupiter for about one month every year. He moves slowly enough that the faster-moving rocky ball we live on easily catches him again after we've left him behind on the other side of the Sun. For the most part, however, Jupiter is easy to see in our sky and so bright among the stars that people often wonder out loud if he is a star. If you watch him for a month and note his position in relation to the stars, though, you'll easily see him progress on his 12-year journey around the ecliptic.

Taking 12 years to go around the Sun, Jupiter spends a year in each one of the 12 zodiac signs and comes into the sky a month later every year. Twelve months, 12 signs, 12 years: I get the feeling that 12 and Jupiter have a lot in common.

From an astronomer's point of view, Jupiter gives us action like nothing else in the night sky. Binoculars alone will clearly show you Galileo's four big moons as they pirouette around this hulking planet. Some people with exceptional eyesight reckon they see the moons with the naked eye, especially when they're at their greatest separation from Jupiter.

Thanks to NASA spacecraft Pioneer 10 and Pioneer 11 in 1973 and 1974, Voyagers 1 and 2 in 1979 and the Galileo space probe that went into orbit around Jupiter in 1995, we have learned a lot about him, his fascinating moons, ever-raging and ever-changing storms and mesmerising bands of cloud.

The Great Red Spot is a storm that has been howling and churning on Jupiter's surface for hundreds of years that we know of. Its fringes are often grazed by smaller storm systems that come and go in the 'shear zones' created as this giant gas ball spins at different speeds at different latitudes.

White clouds of ammonia crystals overlay the red zones that well up from deep within Jupiter. Being a fluid gas planet, Jupiter's equator spins faster than the poles, with the zones in between doing their own independent rotations. The result is the stripy bands of cloud that are easily visible from Earth in even a small telescope.

It is thought Jupiter has a solid metallic hydrogen core, but the planet is otherwise mainly comprised of hydrogen and helium gas. Some people find this hard to fathom because it's easy to think of gas as nothing much. However, just as the Earth's atmosphere has weight that we experience as atmospheric pressure, so too does Jupiter. In fact, this weight gives the planet so much gravity that hydrogen or helium-filled balloons that float away in Earth's atmosphere would sink on Jupiter. Earth doesn't have enough gravity to hold those gases close but Jupiter's gravity is so powerful it clutches all of its hydrogen and helium tightly to it.

Astrologers know Jupiter as the great benefactor; the bringer of good things and opportunity. Astronomers understand the beneficial side of Jupiter too, like its ability to shield the inner solar system from stray rocks and debris that could otherwise smash into Earth. Jupiter has been called the vacuum cleaner of the solar system, as it hoovers up and deflects any asteroids, comets or rocks that stray too close to his enormous gravitational field.

Some of the space flotsam captured by Jupiter has become part of its collection of 79 moons. Other debris has been hurled out of the solar system by Jupiter's incredible forces, while some has been flung towards the Sun on altered orbits. Occasionally Jupiter sucks space rubble into the bosom of the planet itself, as I was lucky enough to witness when Comet Shoemaker-Levy 9 disappeared into its churning gaseous skin in 1994.

Of the 79 moons (at the time of writing), eight were quite likely formed at the same time as Jupiter and have orbits travelling in the same direction as its spin. Four of these moons are smaller than 200 km across and orbit less than 222,000 km from the planet if you measure from its centre. That's pretty close considering our own moon is almost 400,000 km away from the Earth. Jupiter measures 143,000 km across, so the closest of these moons skims just 60,000 km above its cloud tops.

The next four moons are the 'Galileans' discovered by Galileo Galilei in 1610. They are all more than 3000 km in diameter – bigger than Pluto – and three of them are larger than our moon. They are so big that if they orbited the Sun on their own, they'd be regarded as planets in their own right, and if they were further away from the bright glare of Jupiter you'd easily see them with the naked eye.

Three of these giant moons enjoy a 4:2:1 relationship with each other as they orbit Jupiter. While Io goes around Jupiter four times, Europa goes around twice and Ganymede once. This amazing dance of orbital resonance causes these moons to gravitationally bounce off each other and boost each other, as well as create great gas tides on Jupiter itself.

Just slightly bigger than our moon, Io is a volcanically active world heated by gravitational and tidal squeezing. It whizzes around Jupiter in less than two days at only 350,000 km above its cloud tops – a bit closer than our moon is to Earth.

Europa seems to be a gargantuan droplet of water with an ice crust riven by cracks generated by tidal squeezing. The possibility of a liquid ocean beneath the shimmering ice veneer suggests we may yet find life on Europa. This freezing marble runs around Jupiter every 3½ days at 600,000 km out from Jupiter. Io goes around exactly twice in this time, giving Europa and Io a 1:2 resonance.

Ganymede, meanwhile, is the biggest moon in the solar system. It is heftier than Mercury, slightly smaller than Mars and nearly half as big as the Earth. Ganymede orbits Jupiter once every seven days from more than 1 million km away.

Ganymede takes exactly twice as long as Europa and exactly four times as long as Io to go around Jupiter, giving the mega-moon a 1:2 resonance with Europa and a 1:4 resonance with Io. They all pass each other with an offbeat, though regular-as-clockwork, rhythm that can easily make you ponder whether the universe is just a coincidence.

The fourth giant moon, Callisto, is outside of this reso-nance and takes nearly 17 days to orbit Jupiter from just under 2 million km away.

Most of Jupiter's remaining 54 moons range out to 30 mil-lion km from the planet, yet instead of being restricted to the ecliptic plane like Jupiter's main moons, these small irregular moons swarm around the planet in every direction. They travel backwards or sideways to the general trend of things, which suggests they are captured objects that drifted close enough to Jupiter to have their orbits around the Sun disturbed. Now they belong to Jupiter instead.

With Jupiter as an 11 mm grape, the four giant Galilean moons would each be 0.25 to 0.45 mm grains of sand. Io would be 33 mm away from Jupiter, Europa 53 mm away, Ganymede 84 mm and Callisto 148 mm out from Jupiter. Jupiter's most distant moon orbiting the planet from 30 million km away would be a microscopic fleck of dust 2.4 m away from our 11 mm grape.

This mini 'solar system' around Jupiter would be nearly 5 m across in our 1 km wide model of the solar system.

Jupiter, on the edge of the inner solar system, starts a trend followed by the next three planets. They are big, gaseous and fast spinning, and each has a large collection of moons. Yet they are vastly different to each other in appearance, behaviour and in the level of our knowledge about them.

Saturn

Saturn is the furthest object in our solar system that can be seen with the naked eye. Its composition of 94% hydrogen, 6% helium, and a smattering of methane and ammonia, means if you had a big enough bathtub filled with water you could float Saturn in it.

Like Jupiter, Saturn spins very fast on its axis; one rotation every 10 hours. As a result it is noticeably flattened at the poles and bulges at the equator. There is much to love about Saturn, but of course the most remarkable thing about this colossal cosmic bath toy is its perfectly flat, beautifully proportioned band of rings.

In 25 years of hosting star tours, my favourite thing to watch is still when people put their eye to a telescope trained on Saturn and flatly refuse to believe what they see.

'Is it real? That *can't* be real!'

'How can it be so perfect?'

'Seriously, did you put a sticker on the lens?'

If only Galileo could have seen in 1610 what we're capable of seeing today. I'd love to have seen his face.

The rings of Saturn are 280,000 km across, while the planet that sits inside them is 116,000 km in diameter, or nine times the size of Earth. If we could somehow drag Saturn between

Earth and the moon, the rings would extend more than halfway to the moon.

Although their surface area is immense, this band of rings is almost impossibly razor thin, measuring just 10 m in places. They are 99.9% pure water ice, with a tiny sprinkle of dust and rocks, all ranging in size from your fist to the girth of a Mack truck. We have no idea how the rings developed or even when.

Rather than wide bands of differing shades as they appear in a telescope, the bands are made of many millions of tiny, loose ringlets that constantly shift and change.

Saturn is almost as easy to see with the naked eye as Jupiter, although, being twice as far away and slightly smaller, it is not quite as bright. Still, it ranks alongside the brighter stars in the sky and moves slowly amongst them along the ecliptic.

Like Jupiter, we lose sight of Saturn for a month each year as it transitions from the evening western sky to the early morning eastern sky.

Almost 370 years after Galileo first spied the rings, NASA space probe Pioneer 11 gave us our first close-up look, when it passed through them in 1979, testing the waters for Voyager 1 to do the same a year later. Pioneer 11 survived a peppering of impacts from tiny chunks of ice and had a near miss with one of the many moons orbiting within Saturn's ring system.

Voyager 1 and Voyager 2 turned up soon after in 1980 and 1981, giving us even more detailed views of Saturn, and in 2004 the Cassini spacecraft arrived on the scene. By far the most sophisticated mission, Cassini settled into an orbit around Saturn before parachuting a little probe, Huygens, through the thick atmosphere of the Saturnian moon, Titan. Cassini had already completed a couple of flybys of this giant moon to reveal, via radar imaging, large lakes, shorelines, mountain ranges and islands.

There are more than 80 other moons that we know of in orbit around Saturn. Seven – including the giant, atmosphere-shrouded Titan – can easily be seen from Earth in a moderate telescope. Titan is the second-biggest moon in the solar system and the only one with a significant atmosphere, mostly methane. About one-and-a-half times as thick as Earth's atmosphere, Titan's orange haze and other Earth-like qualities have long been a source of fascination for scientists.

Not that Saturn's other moons are dull. One 500 km wide icy moon, Enceladus, shoots geysers of water into plumes that follow it around in its orbit. Other Saturnian moons bounce in and out of the rings, tearing chunks out of them as they go. Some moons thread through gaps in the rings, while others orbit the planet far away from the ring plane.

If you shrink Saturn to a 9 mm grape, the rings are 23 mm across and only microns thick. Titan is a 0.4 mm grain of sand nearly 100 mm away from Saturn. Saturn's family of moons are spread on nearly the same scale as Jupiter's mini 'solar system'.

Jupiter sneaks up and overtakes Saturn once every 20 years in yet another piece of orbital resonance. This is called a Great Conjunction, and in mid-2000 I witnessed this rare manoeuvre with my own eyes as Jupiter and Saturn drew close in the sky. I was able to watch them approach each other, pass each other and then move apart. This watching was a little tricky as they performed this manoeuvre while they were around the other side of the Sun.

By this time in my astronomical career, my awareness had grown to the point where even though I couldn't see Jupiter and Saturn, I knew what they were doing with each other. They did reappear in the early morning sky together a few weeks later to confirm my growing understanding of this momentous event.

As a most auspicious omen for the publishing of this book, Jupiter is approaching Saturn once again as I write. I have watched them, have held them in my conscious awareness in their interaction with each other since that year 2000 Great Conjunction, and now, in 2020, they are doing it again.

As I look out into the evening eastern sky in mid-2020, they're so easily visible that I'm telling people to look for them by saying, 'If you see two bright stars close together in the sky and then wonder if they're Jupiter and Saturn, they will be!'

This Great Conjunction, or passing each other in the sky at only 0.06 degrees apart, is a 21 December 2020 event. They will pass so closely that you could easily block them both out with one finger held at arm's length. If you're reading this after 2020, don't worry, you can watch them drifting apart until they're opposite each other in the sky in 2030, and then watch them come back together by 2040.

Disappearing rings

Taking 29½ years to go around the Sun, Saturn travels a quarter of the way around – or from its own solstice to its own equinox – once every seven years. We can see this transition easily. Saturn, rings and all, is tipped over at 27 degrees in relation to the Sun. This is the same sort of tipping as Earth's 23½ degree tilt. As a result, Saturn shows us the bottom side of its rings for 14 years and the top side for 14 years.

The rings go edge-on from our viewpoint as Saturn goes through its equinox. Seven years later, at Saturn's solstice, we have one of the poles pointed towards us and we see the rings opened out wide to 27 degrees. Seven years later they're edge-on again and seven years after that the other side of the rings is revealed to us at the next solstice.

The rings catch more sunlight when they're opened out, adding to Saturn's brightness around the years of its solstices. Every 14 years when the rings go edge-on, they're very hard to see without the right equipment. It's why Galileo thought the phantom 'moons' he observed in 1610 had disappeared when he looked at Saturn a few years later. I'll never forget the first time I saw the rings at that angle.

I'd been listening to my mate Peter Birch from Perth Observatory talking about it on the radio in 1995. 'Most telescopes are not going to be able to pick up the rings of Saturn at the moment,' he told his audience. 'But Greg Quicke up in Broome has got a 10-inch telescope and he'll be able to see the line of the rings dissecting Saturn's equator.'

I figured I should have a look and, sure enough, there were no 'rings' visible per se, just a line.

While they are definitely special, there's a lot more to love about Saturn than just its rings. Although not as obvious as the storms on Jupiter, there are beautiful cloud formations encircling Saturn with different bands and shading. And, of course, you can see anywhere up to six of its moons in a good telescope.

It's a good thing Saturn is blessed with the other fine features, because in June 2019 NASA reported it had received data from the *Cassini* spacecraft that showed the rings are disappearing – and at a very rapid rate. So get out and look at them while you can, because they're only expected to be there for another 300 million years. I'm just going to wait and see on that one.

Uranus

Although it is four times the size of Earth, Uranus is so far away we simply can't see it with the naked eye. In spite of it being a 'gas giant', Uranus wasn't discovered until 1781 and even today a powerful telescope will only show it as a very dim disc.

The extreme low temperatures on Uranus mean that its gaseous atmosphere is more like a giant slushy. It wasn't until 1986 – 225 years after Uranus was discovered – that Voyager 2 flew past it and showed us a dull bluish world wrapped in a chaotic envelope of high-speed winds and tumultuous storms that rage in its upper atmosphere. Voyager 2 has been our only visitor to this frozen outpost.

Uranus – which has a strange collection of 27 moons, ranging in size from 18 km wide to nearly 1600 km across – spins on its axis once every 17 hours and 14 minutes. Unlike the other planets, however, it is tipped over on its side, so that it seems to roll around the Sun, pointing one pole at the Sun for 42 years, and then the other, as it takes 84 years to make one trip around the Sun. The result is the poles are either bathed in sunlight for about 42 years at a time or shrouded in darkness. From our viewpoint on Earth, Uranus rolls from bottom to top or top to bottom at its equinoxes.

Although they're not as visible as those around Saturn, Uranus is encircled by 13 rings. The inner rings are narrow and dark, while the outer rings are brightly coloured. With a minimum atmospheric temperature of –224°C, Uranus holds the record for the coldest temperatures recorded on any planet in the solar system. This doesn't make it the coldest planet – that title belongs to Neptune. Although Neptune doesn't get as cold as Uranus, it is, on average, colder.

Neptune

Considered a near-twin to Uranus, Neptune is the eighth planet from the Sun. At 17 times the mass of Earth it is slightly more massive than Uranus. Neptune is denser and physically smaller than Uranus because its greater mass causes more gravitational compression of its atmosphere. Neptune goes around the Sun once every 164.8 years at an average distance of 4.5 billion km.

After Voyager 2 flew past Uranus in 1986, the spacecraft continued on to visit Neptune in 1989. The images beamed back to Earth showed an incredibly beautiful blue world surrounded by 13 rocky moons. Ice geysers regularly spew water and ammonia ice into Neptune's upper atmosphere.

As with Uranus, Voyager 2 has been our only visitor to Neptune, so our knowledge about both of them is limited, which is kind of cool. I like a bit of mystery. From a stargazing point of view, however, they are hardly worth the trouble of tracking down. When you do, Uranus is a dull, fuzzy, blue-green disc and Neptune is much the same, only smaller.

Any time I bother to point a telescope at them, people usually react the same way: 'Is that it?'

It's a good thing there's much more to see out there.

Chapter 10

Pluto, Or Not Pluto?

As far as Earth history is concerned, Pluto didn't exist until 1930. Nineteenth-century astronomers had theorised about the existence of a ninth planet after it appeared that the orbits of Neptune and Uranus were being affected by an unknown object within the solar system. They even gave it a cool, theoretical name – Planet X.

In February 1930, after he spent a year scouring space for this mystery world, US astronomer Clyde Tombaugh finally found Pluto moving slowly among a backdrop of fixed stars. It was promptly declared the ninth and most distant planet in our solar system.

From being out there on the edge of eternity, Pluto quickly entered the human imagination, permeated science fiction and became the furthest circle from the Sun in diagrams of the solar system inside millions of upgraded school textbooks the world over. Well done, Clyde Tombaugh! And hooray for Pluto – welcome to the family.

Now get out!

In 2006 – following the discovery of thousands of other bodies beyond Neptune – a vote by the International Astronomical Union (IAU), with four per cent for, three per cent against and 93% abstaining, due to thinking it was a waste of time, decided to strip poor old Pluto of its planetary rank and busted it down to a 'dwarf planet' instead. According to the IAU, only the four rocky worlds of the inner solar system and the gas giants of the outer system are designated planets, making Neptune the furthest-flung world from the Sun.

Pluto was thought to be the biggest object located in what's now known as the Kuiper Belt, a donut-shaped region of debris orbiting the Sun about 3 billion km beyond the orbit of Neptune. However, it seems that Eris, an object three times further than Pluto from the Sun may be very close to the same size, if not bigger. The Kuiper Belt is proving to be a rich hunting ground for many small rocky worlds.

Regardless of its shifting human classification, Pluto has long represented the last unexplored 'planet' in our solar system. All astronomers knew was that it was 5.1 billion km away and smaller than our moon. When Clyde Tombaugh first spied Pluto in a telescope, it appeared as a barely perceptible pinprick of light. Even the mighty Hubble Space Telescope could only manage a fuzzy, heavily pixelated image of the enigmatic erstwhile Planet X.

Unsurprisingly, Pluto has always had a group of dedicated fans inside of NASA. These amazing astro-nerds relentlessly pushed the American space agency to send a machine on a 5 billion km odyssey to check it out. In mid-2015 – just 85 years after Pluto was first discovered – the NASA spacecraft New Horizons made an historic pass of Pluto, following a ground-breaking nine-and-a-half year journey from Earth.

In November 2019, a bunch of these very same NASA guys invited me to join them in Darwin for an 'occultation' event, where one of Jupiter's Trojan asteroids was passing in front of a star. With 26 astronomers and 14 telescopes, we spent three nights practising for this once-in-a-lifetime event only to be thwarted by the seasonal monsoon clouds on the fourth and critical night. We still learned plenty and for me, in particular, hanging around with the group who were responsible for the New Horizons mission to Pluto still gives me goose bumps thinking about it.

The photographic images New Horizons disgorged to us across space blew the minds of every living astronomer. Where we were expecting to see a fairly featureless ball of ice, Pluto was revealed as having a complex and breathtaking terrain. The richly varied surface, of mainly frozen nitrogen, boasts mountain ranges of pure ice that reach 3 km into the Plutonian sky. There are vast ice floes too, and deep chasms that make America's Grand Canyon look like a crack in a footpath.

In an even bigger surprise, New Horizons showed us Pluto has an atmosphere. The blueish hue of the wafer-thin atmospheric skin of nitrogen captured by the spacecraft suggests that if you could stand on Pluto at certain times of its orbit, you'd look up into a blue sky. New Horizons also managed to snap some stunning photos of the giant moon, Charon – the largest of five moons in orbit around Pluto – before zooming further from Earth and deeper into the Kuiper Belt.

We've found other large objects way out there in the boondocks of the Kuiper Belt, including Sedna and eight other known siblings. Classified as a Scattered Disk Object (SDO), or possibly a Distant Detached Object (DDO), Sedna takes 10,000 years to orbit the Sun on an orbit that brings it as close as three times the Pluto distance and taking it out to 25 times the Pluto distance, or nearly 1000 AU.

After all the hoo-ha the IAU went through to kick Pluto out of the solar system, I reckon they're now kicking *themselves*, because it has proven to be an incredibly varied and beautiful world. People often ask during my star shows, 'Why did we kick Pluto out?' Rather than go into the nitty-gritty of the IAU's reasoning, I prefer to settle the matter by holding a vote.

'Who wants to put Pluto back in?' I'll ask.

One hundred or so hands are invariably thrust unanimously into the night sky.

'Whether we just voted Pluto in, or whether those guys voted him out, makes absolutely no difference at all to Pluto,' I'll continue. 'Pluto is still doing what Pluto has always done. The opinions of humans are just not that important.'

To be fair to the IAU, they were faced with a bit of a dilemma. Across 2004 and 2005 three other 'planets' similar to Pluto were discovered; Haumea, Eris and Makemake, all of which joined the ranks of the dwarf planet Ceres that was discovered in 1801. The IAU had to decide, 'Do we call all of these things planets as well, or do we kick Pluto out?'

The rest is history . . . for now.

Oh yeah, and we had to rewrite all the school textbooks. Again.

Like Uranus and Neptune, Pluto is hardly worth worrying about from a stargazer's perspective. Even in a 10-inch telescope it is the smallest speck of light you can see, and telling which one it is in a field of stars is a matter of checking the same stars a few days later to see which one has moved.

But now you know exactly what it is you'd be looking at in all its icy, mountainous and blue-skied glory, maybe you want to try checking out Pluto once in your life just to say you've seen it.

Chapter 11

The Moon

Ever since I first felt the moon tug at my life in the waters of Roebuck Bay, I've been tuning in to its remarkable powers of regulation of life on Earth. For a little old world, the moon packs a powerful punch as it dances with the Sun and the Earth. Most people take this finely balanced celestial dance for granted, but when we start to open our eyes to the ups and downs we might just start to see the crucial role it plays in our lives. It's definitely worth knowing how it all works.

The moon cycle and its manifestation in our lives is complex and there are many factors at play. The moon's orbit around the Earth, its various phases, its effect on the tides and its role in lunar and solar eclipses are all questions that need to be explored to get a rounded understanding of our nearest celestial neighbour.

The moon's orbit

The moon, travelling at 3679 km/h, or 1.022 m/s, in relation to its journey around the Earth, is in a gravitational and energetic relationship with the Earth as they both wind their way through space. In this orbit, the moon goes around the Earth in nearly 27½ days. It also rotates on its axis once in the same time, so, from our perspective on Earth, the moon always shows us the same face. It is locked into what is called a 'synchronous' or a 'captured' orbit. There are only a handful of people in the teeming billions of humanity who have ever laid eyes on the far side of the moon – simply because it always faces away from the Earth.

The first photograph of the far side of the moon was snapped from a Russian spacecraft in 1959. The first human beings to see it with their own eyes were the crew of NASA's Apollo 8 mission in 1968. The last humans to see the far side were those on the last manned mission to the moon, Apollo 17, in 1972.

Although it is sometimes referred to as the 'dark side of the moon', the part of the moon facing away from us experiences just as much sunlight as the side we see.

The Earth travels nearly one 12th of the way, or about 30 degrees, around the Sun during the 27-odd days the moon spends going around the Earth. This means the moon needs an extra couple of days of travelling around the Earth to catch up with and come back to the same relationship point with the Sun and the Earth. In other words, even though it takes 27 and a bit days for the moon to go around the Earth, it takes 29½ days for the moon to go from one full moon to the next.

This is called its synodic period. A synod is a meeting, in this case of the Sun and the moon.

The moon travels around the Earth along the ecliptic, or at least close enough to the ecliptic for us to see it passing all of the slower moving planets once a month.

You've quite likely seen the classic view of a crescent moon in the evening sky with a bright 'star' close beside it. This is when my phone rings off the hook in Broome with people asking, 'What's that star that's always next to the moon?' Well, it's usually Venus playing its eight-month role as the 'Evening Star'.

The moon passes Venus as a two- or three-day-old crescent moon on one particular night for every one of those eight months. If you take a look the following night, the moon and Venus won't be together. Check the moon again the next night and you might wonder where the 'star' has gone and what's going on, just like I did when I first started paying attention. This is the moon going around the Earth.

To circle the Earth in nearly 28 days, the moon moves one 28th – or about 13 degrees – across the sky every day. It spends nearly 2½ days in each zodiac constellation. That's why you can see it alongside Venus one night and it will have moved along in its orbit by the same time the next night. Note its position each night and you will be well on your way to tuning yourself into this fascinating cycle of the moon going around the Earth.

Deception and illusion

At a casual glance the moon can be a tricky thing, almost deceptive in its nature. If you watch it for one night only, you'll think it is travelling from east to west across the sky – just like all of the stars, planets and the Sun appear to do. I'm sure you've seen the moon 'rise' in the east and 'set' in the west. Yes? Of course you have. Except that's not what's going on at all, is it?

The only reason the moon appears to move from east to west and to rise and set is because the Earth is turning. The Earth turns faster than the moon goes around the Earth. So the biggest and most obvious movement you're seeing in the night sky is still the Earth turning.

This continuous movement of the Earth, rolling relentlessly through space, makes it look like the moon is moving the opposite way to how it actually is. For you to see and tune in to the moon's true movement, you need to watch it from one night to the next and then it will become obvious which way the moon is actually going. The moon moves from west to east, the same direction that all of the planets and the Earth travel around the Sun along the ecliptic.

Moon phases

During its 29½-day cycle from new moon to new moon, the moon waxes and wanes, or appears to grow bigger and smaller in our sky. It waxes from being a small crescent to a bigger crescent until it is half lit. From that point the lit moon grows larger, or 'gibbous', until it is full. Slowly the process then reverses and the moon wanes all the way back to a thin crescent in the early morning sky, before disappearing altogether for a few days at the new moon.

The key to figuring out how the phases of the moon work lies in noticing where it actually is in relation to where the

Sun is, as seen from Earth. When the sky is clear you can see the Sun in the sky during the day. Sometimes you can see the moon there too. At other times the moon is there at night and the Sun isn't.

Often I'm asked, 'Isn't the moon in the sky every night?' The answer is 'No'. The moon spends half of its time above our horizon and half below. In other words, the moon spends half of its time around the other side of the planet where you don't see it. The rest of the time it's above your horizon, where you can see it if the sky is clear.

The timing of the moon's appearances is not so much night and day but by this very cycle of the moon that you can become familiar with, simply by looking. Watch it every day for a month and you'll quite likely still be scratching your head. Watch it for another month and you might pick up some of its secrets. Watch it month after month after month and you'll begin to know in the very fabric of your being what is going on.

Stay in front of your TV or computer screen, maybe trying to figure it all out in your head with theory, and you'll be forever in the dark. Step outside and you can ditch the theory in favour of the reality of a real moon in the sky as seen from a real planet below your feet.

As the Earth rotates once a day, the moon stays relatively still in relation to the stars, meaning that you can watch it stay still in relation to Venus or Jupiter or whatever bright star or planet it happens to be visiting that night. It does move half a degree, or about the width of its own diameter, once in an hour, so you'll notice it moving amongst the stars if you're looking for this movement.

Earth's rotation will, at some time in every 24-hour day, turn your part of the planet towards the part of the sky the moon is in, so you'll see it, unless it is the time of new moon. Our rotation will also turn your part of the planet away from the part of the sky the moon is in, so you won't see it.

When we see the moon and *how* we see the moon illustrates the lunar cycle, or the relationship between Earth, Sun and moon. This relationship also determines how much of the moon is lit up for our viewing pleasure. The phases of the moon are at least a little familiar to most people. The cycle is new moon – first quarter moon – full moon – last quarter moon – back to new moon and so on, as it waltzes with us through eternity. Now is as good a time as any to learn the steps.

New moon

This is when the alignment in space is Sun–moon–Earth. The moon is between the Sun and ourselves. It is closer to the Sun than we are. You won't see it at night because it will be in the sky during the day. You won't see it during the day either, because it will be so close to lined up with the Sun that it will be completely lost in the glare of the daytime Sun. In this phase the Sun is shining on the far side of the moon that's facing away from us and the moon's night-time side is facing us. If you look at where the Sun is at the time of new moon, that's where the moon is too.

Three-day-old moon

A few days after passing through the new moon alignment with the Sun, the daytime side of the moon is starting to be turned our way as the moon moves away from the Sun's position and higher into our evening western sky. That's when we see the classic two- or three-day-old crescent moon. Most of the night-time side of the moon is still pointing in our direction and the moon is still closer to the Sun than us.

This is when we see 'the old moon in the new moon's arms'. It's a beautiful spectacle with the lit-up crescent moon and the outline of the rest of the moon visible in much fainter glow. The night-time side of the moon is illuminated by 'Earthshine', or the sunlight that's reflected off the fully lit daytime side of the Earth. Just as a full moon lights up the night landscape on Earth, a full Earth lights up the night landscape of the moon.

A few days after this, the alignment gets even further out of whack and the Earth is no longer as full as it would be seen from the moon. There is simply not enough Earthshine to illuminate the night-time side of the moon. From Earth that unlit part of the moon falls so dark it appears to be missing.

First quarter moon

A week after new moon we see a half-moon in the sky. So why do we call it first quarter? Because the moon has travelled one quarter of its way around the Earth, that's why. At this time, the Earth and the moon are both the same distance from the Sun. We can look across space at the moon and see half of it lit up and half of it completely invisible in the dark. The shadow line is the line between night and day on the moon. The Earth will be in the same state of illumination if you're looking at it from the moon. You'll see half of the daytime side and half of the night-time side. The light from cities on the night-time side of Earth would stand out clearly.

Cast your gaze across space in the direction of the Sun – even if it's below the horizon – and you'll be able to trace the light coming from where the Sun is to shine on the moon. This will help you to see how the moon is being lit and why you are seeing it lit the way that it is.

In Broome and the rest of the tropics at this time, the half-moon is overhead at sunset. This would be more like due south if you're in the US, UK or Europe, or due north if you live deep in the southern hemisphere. No matter where you are, however, if you point one arm at the moon and the other at the Sun, your arms will be near enough at a 90-degree angle to each other, or square to each other.

During the course of the evening of first quarter moon – as the Earth turns and as the Sun appears to sink below the horizon – the Sun and the moon will maintain their distance, and the moon will disappear over the western horizon around midnight. So the first quarter moon appears over the eastern horizon at lunchtime, is high in the sky at sunset and disappears over the western horizon around midnight.

I'm often asked why we sometimes see the moon during the day. Look at the moon and then look at the Sun and you'll quickly realise that both of these worlds just happen to be in the sky together and that the Sun is simply shining across space to light up the moon – even during the day!

In the days following first quarter moon, the moon is further away from the Sun than we are. The angle between the Sun, Earth and moon grows beyond 90 degrees. Your arms, pointing at the Sun and the moon, will be spreading wider.

You see more than half of the moon lit up as it becomes 'gibbous', or fat. It is heading towards the next alignment in space of Sun–Earth–moon.

Full moon

A week after first quarter moon, the moon lines up opposite the Sun in our sky. The Sun shines past the Earth and onto the full face of the moon, giving us the well-known delight of a full moon. Point your arms now and they will be pointing in opposite directions, or at 180 degrees. This is why the full moon 'rises at sunset', is highest at midnight and 'sets' at sunrise. You won't see the moon at all during the day because it will be below the horizon and opposite the Sun all day.

Because the moon's orbit is inclined to the main plane of the ecliptic by 5 degrees, an exact alignment of the moon with the Sun and Earth only happens if the moon is at one of its nodes. More about the nodes later.

The moon misses a direct alignment with the Sun at most new moon and full moon events. We simply see a full moon as it passes to one side or the other of the Earth's shadow or the new moon passes to one or the other side of the Sun without making an eclipse.

Within a day of full moon, the Sun–Earth–moon alignment drifts out of whack and the moon starts to wane, or get smaller. Once again we see less than a full moon as it begins to swing out to the side of the Earth. From then on in, the moon appears over the eastern horizon 50 minutes to an hour later each night. Astronomers love this time as it means there is an extra hour of dark, deep space observing each night.

Last quarter moon

A week after full moon, our little companion world is fully out to the side of the Earth once more – only this time it's on the other side of the planet in relation to the Sun. You can point your Sun and moon arms at 90 degrees once again. In this phase the moon will appear over the eastern horizon around midnight, be overhead at dawn and stay in the sky until lunchtime. This is when you'll easily see it during the day while you're having your morning coffee and croissant.

Waning crescent moon

For the next week, only the after-midnight party-goers and the early morning risers get to see the moon in the pre-dawn dark eastern sky. As it becomes a thin crescent once again, you start to look in the same general direction as the Sun to see the moon. Your arms are going to start closing together day by day, until they're pointing in the same direction. We're coming back towards the initial Sun–moon–Earth alignment and we're looking at the moon's night-time side once again, with only a small sliver of its daytime side lit up as a classic crescent moon.

The lunar cycle

29½ days after we began this journey through the moon phases, it returns to its starting point, aligned in space as Sun–moon–Earth, of the new moon.

Super moons? Really?

The 29½-day lunar orbit brings the moon as close as 360,000 km to Earth and sends it as far away as 420,000 km. This has given rise to a lot of fuss in the media about so-called 'super moons'. When it's on its closest approach to Earth and it's a full moon at the same time, the moon *will* appear slightly bigger. But the difference between the alleged super moon and a regular moon is like comparing a 19-inch pizza with a 20-inch pizza. Can you tell them apart? I can't – and I have looked at the moon *a lot*.

Is the moon upside down?

I get a lot of visitors from the northern hemisphere turning up to Astro Tours and one of the most common questions they ask is: 'How come the moon is upside down?' It's a fair enough inquiry. Since we only ever see one side of the moon, humanity has grown accustomed to the various features of its surface – most notably a dark expanse called the Sea of Tranquility. That's where Neil Armstrong made his historic interplanetary footfall in 1969.

In Australia the Sea of Tranquility is visible on the left-hand side of the moon's surface. In the northern hemisphere, however, this dark patch appears on the right-hand side of the moon. Not only that, the moon appears to be reversed in the northern hemisphere – the shadow of gibbous moon and the glowing fingernail of a crescent moon face the opposite way to southern-hemisphere locations like Australia. Back to front? Left to right? What's going on?

While we all see the same moon regardless of where we are on the planet, it appears to be a different way up because *we* are a different way up. As humanity, we are all standing in different places on this big round rocky ball and all of our feet are pointing at the centre of the Earth. That means that someone standing in Sydney is going to be completely upside down compared to someone standing in London. This astronomy stuff is a great way to mess with your head – or is that your feet?

Is the moon upside down? No. We are.

Stargazing and the moon

The lunar cycle plays a big part in our access to prime stargazing skies. The light of a full moon dominates the night sky and washes out many of the fainter stars. In the days following full moon, however, it appears in the sky an hour or so later every night, giving us an increasing window of time when there's no star-killing moonlight. These are awesome nights for exploring deep space through a telescope; for seeing star clusters, globular star clusters, galaxies and nebulae, or for simply lying back under the magnificence of our own galaxy, the Milky Way.

These deep-space nights last right through until new moon and into the first week after new moon. That gives us a good three weeks of dark, star-filled nights. A new crescent moon hasn't yet built up enough brightness to ruin a dark star-filled sky and the bonus of stargazing in the week after new moon is exploring the moon itself in a telescope. You can see amazing detail in the moon's intriguing craters, its craggy mountains and valleys leading off across lava-flooded plains. It also disappears over the western horizon well before midnight and so allows plenty of civilised hours for a full-on deep-space exploring session.

The week from first quarter moon to full moon is the trickiest for stargazing. The moon itself still looks spectacular, although some of the fainter stars start to vanish in the strong moonlight. In the days leading into the full moon, the Milky Way fades in its enormous glare and it's difficult to find all but the brightest stars.

The good news is you can still clearly see any of the planets that are above the horizon. These 'wanderers' are not at all intimidated by the light of the moon. It's still easy to pick out the moons and the surface detail on Jupiter, the rings of Saturn and the phases of Venus.

So what about lunatics?

Does the moon affect our lives? It's a question I often get when we are out there under a night-time Kimberley sky. Well, you only have to talk with anyone who works in emergency services, nurses, teachers, the police and the booze merchants to know that the moon has an effect on all of us. The crazies come out. And the crazy can come out in us too!

It makes sense when you think about the movement and the interactions between these three very large worlds that are the most significant beings in our lives. One of them is so significant that we are very much a part of it, living on its surface and relying on it for our existence.

Another of them plays an even bigger role in our lives,

being the very source of this world that we live on and everything that goes on upon it both physically and energetically. The third world is the least significant of the trio, although it is big enough and powerful enough to be the regulator of our life here on Earth, whether we have an awareness of this or not. Most of us don't. Most of us are lucky to think about what is going on 10 feet around us, let alone 150 million km away to the Sun or 400,000 km to the moon, or even 12,756 km away to the other side of the Earth.

Of those three, the Sun is the source of everything that we are. I mean this literally, with the substance of the Sun being the same substance that we are made of, with the energy of the Sun, the heat, the light and all of the other energies that we receive from the Sun every day, many of which we are still yet to detect or discover, driving every process here on Earth, including – I'd suggest – the very words that are appearing on this page as I write.

The Earth is our stage, the place where we do our very best to be everything that we came here to be, the place of opportunity, of challenge, of hardship, of love and of joy. What happens on Earth happens to us.

Often we think within our 10-foot radius that everything is happening to us. Step outside of that 10-foot radius and you'll frequently see that the same things are happening to others around you at the same time. Perhaps that can make it easier if we're able to see that it is not just all about us.

The moon's interaction with the Earth and the Sun is a case in point. This little old decrepit ball of black, seemingly lifeless rock works the tides in a very blatant and obvious way. It might be worth considering that its effect on us is equally blatant and obvious, if we'll only take notice. How many times have you been knocked flat, only to realise that it is full moon time, perhaps after it's all over and you've picked yourself up off the floor? There are not many here on Earth who are immune to this cycle. There are, however, those who've learned to work with it rather than being worked by it.

If we think about the Sun as our source and about the moon as a potentially disrupting influence, and then think about the way the moon comes between us and our source and then gets out of the way from our source in a regular ongoing cycle through the ages, it might be worth observing what happens at different stages of this cycle.

At new moon, the moon is between us and our source, potentially cutting off or resetting the cycle, just like when you switch the computer off and then back on again. The power is off, the energy is low. However, it is also the germinating of a seed, the seed of the new lunar cycle with all of the fresh start that this implies. As the moon moves out of alignment and allows the Sun to shine through to the Earth, perhaps you can feel the building of that new energy. As the moon pushes towards full moon, the pressure is on and the first cracks are appearing in the demeanour of those who are inclined that way.

There are others on the planet who are starting to paddle their surfboards about now, looking over their shoulders at the building swell and getting ready to jump to their feet and ride the wave of energy delivered to Earth at every full moon.

Every major religious festival on Earth recognises this time of power. Easter is a full moon festival set by the first full moon following the March equinox. The eastern festival of the Buddha is set by the following full moon, with many other cultures on

Earth using and recognising the energy uniquely available at this high time of the lunar cycle.

Indeed, despite the western world abandoning the lunar calendar in favour of the solar-based Gregorian calendar in 1582, the lunar calendar is still in use throughout much of the eastern world. Even though the Gregorian calendar still uses months to give us the illusion of being related to the moon, it has long since stopped having any relevance in this regard. It is almost like there was a plan to disconnect us from this natural cycle. I guess this gives us the option to consciously engage with it, rather than having it imposed on us by the traditions of the eastern and Hebrew lunar-regulated life.

Taking a fresh look at the significance of the regulatory rhythm, the low and the high energy times of the lunar cycle might be worth some consideration. So what is going on? At new moon, the moon is blocking our access to the source of everything that we are. At full moon the moon is as far out of the way as it can get. It is opposite the Sun in our sky, giving the Earth uninterrupted access to our source and we receive the full blast of energy from the Sun.

So is it the moon that sends everyone crazy at the full moon time or is it a full blast of solar energy that is otherwise coloured by the dulling-down influence of the moon getting in the way of and interfering with the very source of all that we are? I'll leave it up to you how you want to think about it.

The moon is a part of our lives and, if you tune in to it, you might find it useful to work with every part of the lunar cycle. If you just want to enjoy it as a stargazer, my advice is simple: when the moon is there, check it out. It's amazing! When it's not there, there is a whole universe to explore.

Chapter 12

At One with the Sun

We live on a planet that we call Earth. We have a bunch of sibling planets in our solar system. At the centre of our solar system is an unimaginable source of power and light, a star that is known by many names, such as Ra, Sol, Helios, or simply the Sun.

It is so powerful that gazing upon it is painful and not something that anyone does for more than a fraction of a second at a time. It is pretty hard to miss: 99.8% of the solar system's mass is contained within this 1.39 million km wide orb of blinding light. The Sun is so immense that our planet could fit inside it 1.3 million times and we would have to line up 109 Earths to match its girth.

And that's just the physical Sun we can see and measure. We often think of it as a perfect golden ball hanging somewhere up in the sky, a long way away and just doing its thing. The truth is we are very much *within* the Sun and not separate from it at all.

Essentially everything in our solar system belongs to the Sun. All of the planets, all of the asteroids, the comets, the trees, the waves, eagles, volcanoes, moons, ants, rainbows, dolphins, motorbikes and human beings are intimately connected to, and made from, the very matter and energy of the Sun.

The Sun is literally the source of everything in this world. *Everything*. All of the matter and energy, and all of the inter-actions between matter and energy that we perceive as activity, have a local source in our Sun. We bathe in the very substance of our nearest star as its solar wind streams past and through us every day. We literally see this substance as light, feel it as heat and we grow with it and measure it in many different ways.

The fact we can perceive the Sun and ponder the inter-actions it causes implies consciousness on our part. I reckon the Sun would be a worthy place to look for the origin of that consciousness, because without the Sun there is no human thought; everything we know on this planet simply would not exist. It's no wonder many past civilisations worshipped it as a god.

Our understanding of the source of the Sun's energy is an evolving one, with many enigmas about it still far from being solved. Our knowledge is relatively crude compared to what is likely yet to be discovered. (Remember, it was only 100 years ago that we thought our pinprick of a galaxy was the entire universe.)

Humankind has thrown up plenty of ideas about the Sun's engine room over the years. Around 450 BC a Greek philos-opher named Anaxagoras taught people that the Sun shines because it is a red hot stone. A couple of thousand years later a German thinker named Julius von Mayer did some calculations that showed if the Sun was a massive lump of flaming coal, it would have burned out long ago.

Next, the spectroscopic discovery of the Sun's hydrogen make-up in the 19th century led to speculation it was a ball of burning hydrogen gas. Once again, calculations showed this model would have run out of fuel by now.

Albert Einstein, with his concept of mass-energy equivalence – $E = mc^2$ – proposed the idea of nuclear fusion that gradually gained favour in the first half of the 20th century and is strongly promoted to this day. The current mainstream understanding is that chain-reaction nuclear fusion is the basic driving process of the Sun and of the stars.

This model has problems too. With the 'campfire nuclear furnace at the centre of the Sun' model, we'd expect the centre of the Sun to be the hottest point and that things would cool down the further we got from that centre. It's the same as taking a step back from your own campfire when it gets too hot.

We measure the surface of the Sun as about 5500°C. Stepping away from the surface, we'd expect it to be cooler. However, the Sun's corona, or atmosphere, or aura, stretching out to more than a million km into space, is more like 2 million degrees. That's like stepping away from your campfire and getting hotter!

Streaming away from the corona in every direction is the solar wind, a stream of charged particles that varies in density and speed depending on the activity of the Sun. With the 'gravity squashing everything so hard that it bursts into a constantly exploding nuclear bomb' model, we would expect these charged particles to slow down once they're ejected from the Sun, as the mighty gravity of the Sun pulled on them.

Instead, what we find is that not only does the stream of charged particles not slow down, it actually accelerates away from the Sun, getting faster and faster the further it goes. The only known phenomenon that will accelerate a charged particle is an electrical field.

The magnetic nature of the Sun and the planets and the accelerating charged particles of the solar wind were all only discovered in the modern space age with satellite-borne instruments. Now we're discovering magnetic fields on every scale in the universe, from the magnetic sheath of our own planet to the magnetosphere of Jupiter, which, if it was visible, would make Jupiter appear to be two or three times bigger than the Sun or the moon in our sky. Our Sun's magnetosphere extends well beyond the planets.

Plasma, the fourth state of matter

We now know that everything in space – that is, everything above a layer of our atmosphere called the ionosphere – is in the plasma state. Plasma is the fourth state of matter.

We're familiar with the three states of matter common here on Earth. They are solid, liquid and gas. In a solid, the atoms and molecules making up a substance are locked into a matrix to form the solids that we're so familiar with. Ice is a solid. In a liquid, those very same atoms or molecules are sliding around on each other to allow that water to slosh about in your cup. In a gas, those same particles are further separated until they move freely and independent of each other. If we contain these free particles, they bounce off each other to give us 'pressure' in a gas.

In a plasma, these very same atoms or molecules are further separated into their charged elements – into their positively charged protons or atomic nuclei and into their negatively charged electrons or negative ions. These charged elements have the ability to organise themselves into currents and into layers. We see this replicated in plasma laboratories here on Earth.

We see plasma on Earth as the flame of your campfire, or in your fluorescent light tube where the gas is 'excited' or raised to the plasma state with the flick of the electricity switch so that it glows. We see plasma with an arc welder where it's in the even more excited and energetic 'arc' mode. Lightning touching down on Earth – with its associated upper atmospheric phenomenon of sprites – gives another powerful example of arc mode plasma connecting the Earth and the sky.

A plasma is an electrically conductive and magnetically active medium. The European Space Agency's Herschel and Planck missions are revealing huge filamentary plasma structures embedded within magnetic fields. We see stars strung along these filaments, connected with each other, just as we are with them.

A stream of charged particles (plasma) constitutes an electric current and there is a growing body of evidence that electricity and magnetism play at least as big a role in celestial mechanics as the force of gravity that has dominated astrophysical thought since the time of Isaac Newton's discovery in 1687.

With $E = mc^2$ teaching us that matter and energy are the same thing, you could think of the solar system as a big electromagnetic soup, with the planets in our solar system as concentrations of this energy/matter within the broth. Their movements stir the soup pot and constantly change the dynamics within the solar system. With their idiosyncratic orbits and resonances and harmonics, each one of the planets plays a role in this mighty merry-go-round.

The planets generate tides on the Sun. This is particularly noticeable when the two biggest planets, Jupiter and Saturn, are on the same side of the Sun as each other. When Jupiter passes Saturn on the inside lane once every 20 years, the centre of gravity, or the barycentre, of the solar system is actually shifted outside the surface of the Sun, with the Sun wobbling around this centre too. All of the planets are in on this act, constantly interacting with each other gravitationally, energetically and physically.

So, it is quite correct to say that our Sun *is* the solar system and that our solar system is the Sun. There really is no separation. In fact, the Sun's influence and the very substance of it is thought to stretch all the way out to the influence of the next sun or solar system. As we are seeing with the work of Herschel and Planck, streams of energetic material connect the stars with each other. Our Sun forms a bubble that stretches to the point where it no longer has the energy to push against the space between the stars, or to the point where it bumps up against the bubble of the next star.

Everything that is within that bubble is a part of the one organism that we call the solar system. The Sun touches and indeed *is* all of it. You and I in our human bodies on our little planet 150 million km away are a part of the Sun. We are made of the same stuff.

Aurora

Solar wind is the name we give to the charged particles that are the very substance and energy of the Sun. It constantly streams past the Earth at speeds between 250 and 750 km per second, depending on what is going on in our solar environment at the time.

We see the solar wind's interaction with our planet through the auroras, the Northern and Southern Lights. These mesmerising light shows occur around the Earth's magnetic poles when the charged particles of the solar wind cause disturbances in the Earth's magnetic field. Energetic particles interact with and strip electrons from the vulnerable and already excited atoms where the magnetic field lines converge at the poles, causing them to glow in ionised light. This is another example of plasma in the glow mode.

Auroras are particularly active when the Sun is particularly active with sunspots or solar storms. When these storms erupt in the general direction of Earth, we are buffeted by a stronger, more energetic solar wind.

A highly charged solar wind can interact with big man-made structures on Earth too, such as power grids and pipelines. It can induce a charge that has the ability to knock out power systems, as it did to the entire province of Quebec in Canada in 1989 following a massive solar storm.

We live within this fluctuating charge from the Sun every day of our lives in the electromagnetic environment of the universe. Luckily, or perhaps by design, our atmosphere and our magnetic field shield us from the more dramatic effects of the solar wind. Otherwise we would be pretty much fried by the incredibly potent radiations of the Sun.

While it is everything to us here on Earth, and the source of all that we are and all that we know, our Sun is only one minuscule piece among the hundreds of billions of pieces in the giant jigsaw puzzle that is the Milky Way galaxy. The Milky Way is an even smaller piece in the grander jigsaw puzzle of the universe.

Chapter 13

Around the Sun

Ask someone to describe a year passing by and they might begin with New Year's Day and proceed to catalogue a series of earthly events; some achievements at work maybe, a personal milestone, a change in a relationship, a birthday party and a holiday or two. Maybe they'd cite the shifting seasons as summer fades into winter and the inevitable return of spring, before coming full circle back to fireworks on New Year's Eve.

In my experience, not many people describe a year as travelling a full circle around the Sun in a 490 million km mega-voyage, but that's precisely what a year is. What's more, you can watch it happen.

It took me years of staring up at the stars to realise the Earth was not only spinning on its axis once a day, it was also hurtling around the Sun. Okay, I knew it on an intellectual level but I certainly didn't know how to see it for myself.

Lying back in my swag over a number of years, I eventually started to notice the same stars would appear in the same part

of the sky at the same time each year. I watched this precision recalibration take place four or five times before the penny
dropped. Four or five times equals four or five years! When it
finally registered that the same stars reappeared at the same
time each year because the Earth had simply come back to
the same place again in space, I almost fell over.

That was the moment I started to *feel* the movement of the
Earth going around the Sun. That's the moment it became real
for me. Even though it's not the sort of thing I feel every day
when I'm walking down the street, when I stop and tune into it
I have a very real sense of this journey that we're taking around
the Sun.

I'm going to suggest that you can too. If you simply start
watching some of the things I'm setting you up to see, you'll
eventually feel this movement for yourself.

It's as fascinating to observe the Earth orbit the Sun as it is to
see the planet turn once a day. It's not something you're going
to see in a couple of hours of observing; however, if you start
watching now, it will really only take you a week. That week
will turn into another week and then a month and then a year,
and then you'll know and it'll be a part of your awareness to be
a space traveller on planet Earth.

It's a matter of tuning in to the Earth and to the Sun. If you're
reading this in the daytime have a look for the Sun now. Bring
it into your awareness. Even if it's cloudy you can get some idea
of its position. If it's sunset, it's in the west. If it's sunrise, it's in
the east. If it's night-time it's below your feet on the other side
of the planet.

Next, tune in to the Earth. It's right there! Give it a little
tap with your foot. Now have a look at the Sun while being

mindful you are watching it from Earth. The Earth is moving sideways in relation to the Sun. That's just how an orbit works. I could get technical and say the Earth is moving at a tangent, or at 90 degrees from the Sun's position, but it's the same thing as saying we are moving sideways. This sideways movement takes us on a huge curve that whisks us around the Sun at 107,000 km/h.

Next, we're going to finetune this 'sideways' so you'll be pointing the right way. Face the Sun and put your arms out wide and you'll be pointing sideways or at 90 degrees away from the Sun. That 90 degrees could be up, down or to either side, depending on the time of day or night and your location on Earth. Let's have a look at how it works at different times of the day.

Going east

At sunset the Earth is dropping away under your feet to go down to go around the Sun. It is moving at 90 degrees from, or sideways to, the Sun's position over there in the west. There will be some left or right in your downwards here too, depending on how far north or south you are, and depending on the orientation of the ecliptic on that particular day at that particular time. 'Down' is still the key here.

By midnight the Earth has turned you 90 degrees, so the Earth is hurtling you through space going east. The Sun is below your feet and you're still going sideways in relation to it. Again, the only thing that has changed here is that you've turned a quarter of a rotation on the Earth to alter your orientation to the Sun.

Going west

At sunrise, because you're now upside down compared to where you were at sunset, the Earth is taking you headfirst to go around the Sun.

At midday you're going west.

The journey is the same. The Earth is hurtling through space to go around the Sun at 107,000 km/h. Our perspective on that direction changes simply because we are turning around once a day on the Earth while it carries us on this bigger journey.

Watching the journey

If we choose to look for this movement every evening – instead of at any other time of the day or night – we can take out the complication that the spinning Earth brings to watching the planet go around the Sun.

The far side of the Sun

So if we're standing and watching the sunset, the Earth is dropping away under our feet at 107,000 km/h. This is real. The Earth is going to take us down and around the Sun to be all the way over there on the other side of the Sun in six months. So not only is six months a different 'time', it is also a different 'place'. It's a place that is 300 million km away from where you are now.

Use your imagination to go there, because that's where you're going. We'll be there in six months' time. It's where we were six months ago. It is a very real place that is hundreds of millions of kilometres away from the part of the universe you're in right now! Have a think about that in six months' 'time'.

On the six-month return journey the Earth will travel back across the ecliptic and through the same set of stars you can see at sunset tonight. It will travel through the same very specific part of the sky at the same very specific time each year to reach the same very specific location you are in right now. And that's what a year is! The Earth going around the Sun once.

A week

You won't notice this movement by watching for an hour or so. To *really* see the Earth zooming around the Sun, you'll need about a week. After the first week, the second one will be easy. Do a month and you'll be on your way to watching the Earth go around the Sun for the rest of your life. Watch it for a year, and then another year, and you may start to feel the whole 'Oh my god, we're travelling through space' thing, to the point you'll know you're on an amazing journey that you're playing a part in.

How do you see it?

To set you up for this, I'm going to remind you of a couple of things you already know:

* The Earth goes around the Sun once every 365 days (or 365.256 days, to be precise)
* And . . . there are 360 degrees in a circle.

1 degree a day

That means our planet moves about 1 degree around the Sun every day. So, if you watch the sky at sunset for two nights in a row, the Earth will have dropped away under your feet and travelled 1 degree around the Sun by the second night.

Because the Earth moved by 1 degree around the Sun, all the stars will appear to have shifted west across the sky by 1 degree. Now, 1 degree is pretty hard to notice.

Four minutes a day

This 1 degree movement also means the stars will appear in their same positions in the sky four minutes earlier the following day. Four minutes a day is pretty hard to notice too!

28 minutes a week

Four minutes a day, though, adds up to nearly half an hour a week. That means any stars in the eastern sky will be in the same position half an hour earlier next week. Stars will set over your western horizon half an hour earlier each week. The Southern Cross will achieve the same angle in the sky half an hour earlier each week.

7 degrees a week

If you watch the same stars in the east at the same time of night next week, they will be 7 degrees higher in the sky. The stars in the western sky will be 7 degrees lower. In a week, the Southern

Cross will appear to have moved around its circle and changed its angle by 7 degrees in the same direction as the clock.

The northern hemisphere's Big Dipper, or Great Bear, will have changed its angle by 7 degrees in the opposite direction as the clock. Remember, though, it's only the opposite direction because you'll be facing the other way. The Earth will still be turning you the same way and the Earth will still be taking you around the Sun the same way. What you do and the way you face affects your view.

Astronomy is often presented as that stuff that is 'out there'. What I like to remind you of is that you're a part of it, and that what you do, and how you do it, and where you go, is a big part of your experience in the universe.

The Earth is going around the Sun: three months

Watch the sky for three months at the same time each night and the same stars you started with on the eastern horizon will be overhead, or 90 degrees across the sky. This is because the Earth goes 90 degrees, or a quarter of the way around the Sun in three months.

Six months

In six months the Earth goes halfway around the Sun, so all the stars we started with on the eastern horizon will be on the western horizon. There will be a completely different set of stars in the sky. That is the Earth going around the Sun!

This is the difference between the stars in the winter sky and those in the summer sky. This happens because you're looking out into space in a different direction because we're completely around the other side of the Sun.

People often come to Broome to escape the cold southern winter to come stargazing with me. One of the questions I frequently get is, 'Why can't we see the "saucepan" or Orion in Broome?' My answer is that they can't see it at home at the moment either. Often their reply is, 'Oh no, we can always see it at home.'

That lets me know that they only ever watch the sky in summer! Which makes sense, when going outside on a cold southern winter's night means clouds and rain. Some of those crisp clear nights in the south are spectacular for stargazing, although it might be so cold that taking the time to look for the saucepan might not even enter your mind.

What's happening is that the Earth is around the other side of the Sun and Orion is simply not in the evening sky. This is true for anyone who is on Earth. No-one can see Orion in the evening sky in the middle part of the year. Everyone on Earth sees the same stars as everyone else on Earth at the same time. The only difference is that these same stars will be shifted to the north or to the south depending on where you are. There will be a big overlap in the middle that everyone can see and there will be parts you can see that someone in the other hemisphere can't, and vice versa.

Everybody who is on Earth, is on Earth

We're all on this planet together and we're all in the same part of space together, and we all have the same view out into space at the same time as each other. Orion, or the saucepan, not to be confused with the Big Dipper – a different constellation altogether – is a summer or Christmas time of year constellation. It is opposite the Scorpio constellation, so you'll always get one or the other in the sky. Just as Orion dominates our Christmas sky, Scorpio dominates the middle part of the year. By noticing this, you're watching the Earth go around the Sun!

A year of changes

One of the great things about our journey around the Sun is that we get a new set of stars to look at every month. The stars coming over the eastern horizon have been out of sight for six months. At least, this is true if you're only watching during the civilised hours of the evening. If you stay up all night, you can get nearly all of them at any time of the year.

Some of these stars are like old friends to me and I look forward to their return each year, while there are still plenty of stars that I'm yet to learn. When they turn up on the eastern horizon each year, I reach for the star chart to remind me of their names and character. There are others still that pass across the sky with only a casual glance from me; relative strangers I've yet to properly meet.

This changing sky during the year as the Earth goes around the Sun keeps me interested, keeps me learning, it keeps me on my toes, and stimulates me enough to keep doing what I do with groups of people under the stars year after year. If I had to watch the same stars every night, I would have long since found something else to do.

Chapter 14

My Own Rising 'Star'

In 2006 I bought myself a motorbike. Not an unusual move for me – I've owned plenty of them – but this one happened to be for sale on the other side of the country in Brisbane. This might seem impractical for a bloke who lives in Broome, but there was method to the madness: the purchase promised me a long, adventurous ride to get it back home.

I took my time on the return journey and spent three months exploring the lush valleys and funky towns of the NSW north coast. There was, however, an itch I wanted to scratch. For a long time I had yearned to get a close-up look at the nation's most famed piece of stargazing equipment – the Anglo-Australian Telescope (AAT).

While I'm still quite fond of my 10-inch Dobsonian, and probably always will be, I've been head-over-heels in love with the AAT ever since I first read about it as a boy. And now it was just a few hundred kilometres away. I turned the bike inland

and headed for the town of Coonabarabran in north-western New South Wales.

The Siding Spring Observatory is perched on the forest-fringed rim of an extinct supervolcano in the Warrumbungle Range west of Coonabarabran. Although about 50 of Australia's largest telescopes are dotted across the mountaintop that reaches more than 1 km above sea level, the 3.9 m AAT is the big daddy of them all.

Completed in 1974 for the Research School of Astronomy & Astrophysics at the Australian National University, the AAT was crafted in an era when things were beautifully constructed. Today we build enormous telescopes that look more like industrial machines, but the AAT is as much a work of art as it is an elegant example of precision engineering. The confluence of form and function in that telescope is truly something to behold. Even the paint job is beautiful. Told you I loved it.

Nowadays there are 10 m telescopes placed on some of the world's highest mountains, but back in the 1970s the AAT was among the world's largest. Although not that big on today's scale, it is still a serious piece of equipment. The kind of innovation astronomers and astrophysicists are able to conjure by adding different instruments to the AAT – and the type of work they're doing there – is world leading.

Siding Spring also happened to be home to one of my heroes – its British-born Astronomer in Charge, Professor Fred Watson. I had been listening to Fred's regular guest spot on ABC Radio for years and was always enthralled and educated in equal measure, whenever he held forth on all things cosmic.

Naturally I was pretty excited as I roared up the winding road to the top of the mountain and the gleaming 50 m high white dome of the AAT came into view. However, I don't know whether or not Fred Watson was there that day, and I was in no danger of finding out.

'Hello, my name is Greg Quicke and I'm an astronomer from Broome,' I enthusiastically told the man in the reception area. 'I was wondering if I could get a look at the Anglo-Australian Telescope?'

It turned out my modest credentials of running star tours in a bush lot out in the never-never didn't have much currency on this particular outcrop of hallowed astronomical ground.

'Um, yeah,' the man replied. 'There's a window over there you can go and have a look through if you want.' Although my heart sank a little with his casual dismissal, I quickly understood that not just anyone can rock up unannounced on a motor-bike and expect a close-up guided tour of one of Australia's top pieces of scientific equipment.

As I rode away on my long haul back to Broome, I never dreamed I'd one day get a chance to experience just that.

A few years later the phone rang at the Astro Tours office (i.e. the day bed on my front porch) and I was stunned when the caller introduced himself as none other than Professor Fred Watson! Fred was semi-retired by that time and he and his good friend and colleague Dr David Malin had started taking groups of interested people around the world on astronomy tours. The eminent gents had a favour to ask.

David, a British-Australian astronomer and photographer, was also something of a hero of mine. Although he's most renowned for the spectacular, ground-breaking colour images of astronomical objects he took using the AAT, David also famously discovered a massive spiral galaxy back in 1986. Three times the diameter of our Milky Way and 1.19 billion light years away from Earth, the galaxy was officially named Malin 1, which *really* put David's name up in lights for all time.

Fred explained over the phone that he and David were planning to bring a group tour to the Kimberley and they were wondering if they might be able to use some of my telescopes – which was totally fine with me.

While they'd made it to the Kimberley and to Broome, unfortunately the logistics of their trip didn't quite work out and we were unable to link up with their group. Furthermore, their plane back home was delayed. That's when I got another phone call from Fred: 'We're stuck here at Broome Airport for half a day, Greg. Do you want to come down, say g'day and have a chat?'

As soon as I pulled up a chair in the departure lounge and met Fred and David for the first time, I knew I was among kindred spirits. We all got on like a house on fire and Fred was generous and effusive when I hit him with a thousand questions about the Anglo-Australian Telescope.

Both men were a pleasure to talk to and a wealth of information about the technical side of astronomy and the art of imaging objects deep in space. As two of the world's leading astronomers were finally about to board the plane, I unashamedly hit the good professor up for a favour: 'Fred, when I come over to the east coast next, can you show me the telescope?'

'Absolutely,' he said, laughing. 'I'll show you all over it. We can work it together and I'll demonstrate all sorts of things it can do.'

I farewelled my new mates at the departure gate and went home to prepare my gear for that night's Astro Tours show. I felt pretty chuffed that a university dropout cum pearl diver cum bush mechanic cum bush astronomer had been so readily accepted and welcomed by what I considered to be astro royalty.

Over the next few years I kept doing my thing, talking to audiences and showing them the wonders of the night sky in the glorious Kimberley. I didn't quite get the chance to drop everything and head over to Siding Spring and take Fred up on his offer but, somehow, the universe seemed to have a way of working these things out.

A lot of international visitors come to Broome and in 2016, unbeknown to me, one such guest at Astro Tours was a TV producer, who subsequently went home to his boss, a fellow named Paul King, a series producer with BBC Television in England. The broadcaster, he explained, was planning to shoot the next series of the popular British astronomy show, *Stargazing Live*, in Australia. In its seventh season and hosted by Professor Brian Cox – the world-famous particle physicist from the University of Manchester – the show was designed to introduce the masses to the wonders of astronomy through interviews, demonstrations, discussion and live observations of the cosmic phenomena playing out above our heads.

Paul King reckoned my practical approach to navigating the night skies would fit in nicely alongside the more technical and mathematical approach taken by Professor Cox. Before I knew it, I was plucked from the ochre sands of Broome and plonked onto the world stage – which just happened to be on the rim of a certain extinct volcano in NSW.

*

When I returned to the Siding Spring Observatory in 2017, along with the cast and crew of the BBC's *Stargazing Live*, I wasn't waved away by the reception staff. Instead, I was welcomed like a bit of a rock star.

The show was to be filmed at the observatory over three nights in March in order to take advantage of the equinox and marry the last 45 minutes of darkness in Australia with an 8 pm broadcast time in the UK. We'd been invited to stay at the facility's brand new 'observer's lodge' and when I was shown to my room I noticed my name was on the door. A little later, Paul King collared me. 'Greg,' he said, 'did you see your name on the door?'

'Yeah,' I replied, thinking nothing of it.

'Did you see how many other names are on doors?' Paul pressed.

'Ahh, nup,' I answered.

'Well, there's only four doors with names on them,' he said. 'You should go back and have another look.'

The other rooms with names on the doors belonged to Brian Cox, his famous co-presenters Dara Ó Briain and Liz Bonnin, and me. I certainly wasn't in the league of Dr David Malin, with his own galaxy named in his honour, but I did feel a little like I had my name up in lights.

The next day, as I was walking down the stairs of the lodge, the lovely Fred Watson was walking up them. I was stoked to see him again and smiled widely, but before I could even open my mouth to say g'day, he beamed back at me and said, 'Yes! I remember!' In other words, he hadn't forgotten his pledge to show me over the Anglo-Australian Telescope. I felt like I'd died and gone to heaven.

Of course, I'd never met Brian Cox or the other presenters before, so I didn't know them from a bar of soap. Since we were all staying together, it didn't take too long to get acquainted. Brian was preparing a presentation about the Milky Way and in the small hours of the early morning I found him at the communal kitchen table deep in thought and mired in complex calculations.

He didn't know me from Adam either, so I just wandered over to the table and said, 'Oh yeah, what are you doing?'

'Oh, hi,' he said. 'I'm just trying to figure out this angle between the axis of the Earth and the Milky Way.'

'Yeah?' I said. 'Come outside and I'll show you!'

'Oh, no, no, no – I've got to calculate all this stuff on here,' he replied, pointing at his laptop. 'Can you see what I'm trying to do here with these calculations? Do you think I've got it right?'

'Yes, that looks about right,' I answered, 'but come outside and I'll show you the angle.'

I eventually persuaded Professor Cox to come outside with me. As we stood in the pre-dawn blackness, I stuck one arm out in the direction of the south celestial pole and pointed the other at the Milky Way.

'See the angle between my arms?' I said. 'That's the angle you're looking for.'

Brian looked at me in silence for a moment, with my arms splayed out at the night sky, before letting out a satisfying 'Ahhhh!' as the simplicity of the demonstration hit home for him. After that I think he recognised I had some good insights to share.

Soon after that encounter I discovered that Brian had never seen the Omega Centauri globular cluster. For my money it is one of *the* most amazing objects in the entire sky – the most radical place you can imagine.

When I started Astro Tours 25 years ago I used to tell people Omega Centauri had one million stars in it. Then I read an article that said it had four million stars, so I started telling people it had four million stars. But up in Siding Spring Brian told me the current thinking is Omega Centauri is home to some 10 million stars. Astronomy is a bit like that. Imagine a compact ball in the sky made up of 10 million stars! And yet in six years of presenting *Stargazing Live*, he'd never once seen it with his own eyes.

To be fair, Brian lives in the northern hemisphere where Omega Centauri isn't visible. He had been to the southern hemisphere plenty of times, so I was surprised nobody had bothered to show it to him. I wasn't going to let that continue.

Since all of the telescopes on that mountain are extremely technical machines for high-grade scientific observation, very few of the operators actually put their face to an eyepiece and peer through the glass of the telescopes themselves. More often than not, they look at incoming data on monitors and use various spectrum analysers and computer models. Not surprisingly, however, some of the terrific people who work there have more traditional, yet still pretty serious, telescopes in their backyards.

A couple of nights later one of the operators invited a bunch of us to his place at the bottom of the mountain to have a look through a 14-inch Schmidt-Cassegrain telescope. There were three or four observation domes on his property, which people paid him to put telescopes in so they could operate them remotely and robotically.

Fortunately the big Cassegrain wasn't being used by the client that night so I was invited to pile into the small dome with an international who's who of astronomy and astrophysics – Professor Brian Cox, Professor Fred Watson, Dr David Malin, Dr Chris Lintott from Oxford University and Professor Lisa Harvey-Smith from the University of NSW, and Paul King, Helen Thomas and Duncan Williamson, who were the BBC series producers.

Since the big Cassegrain was normally robotically controlled, it didn't have a finder scope attached. This made it very difficult and awkward to use, and none of the people with me knew how to operate the thing properly. I was so determined to give Brian Cox his first-ever look at Omega Centauri, though, that I started manhandling the unwieldy big scope into position to try to locate the globular cluster by hand and eye.

After a bit of searching around where I knew it should be, I finally zeroed in on Omega Centauri glistening in space nearly 16,000 light years away. 'Okay, got it,' I said, being sure to keep the telescope steady and the ball of 10 million stars in the centre of the eyepiece. 'Now, lock it off.'

After a couple of the others twisted the two locking devices into place, I ushered Brian over to take a look. The Cassegrain is a research-grade telescope and, since we were in the middle of nowhere on a moonless night and at 1100 m elevation, Omega Centauri looked more stunning that night than I had ever seen it before. The expression of awe and wonder that came across Brian's face when he looked into the eyepiece said more than any words could. I could see his mind was completely blown.

That episode let me show my strengths and underscored the

value I hoped to bring to a show like *Stargazing Live*. Following the success of the BBC series, we filmed an Australian version of *Stargazing Live* for ABC TV in April 2017. It was co-hosted by Brian and Julia Zemiro, and once again I was in the practical astronomer seat to give demonstrations of the night sky and to explain some of the practicalities of how the cosmos is set out. A second Australian three-night series aired on ABC TV in May 2018, and again when Brian and I presented *Stargazing: Moon and Beyond* in July 2019, celebrating 50 years since the moon landings. By then I had well and truly become known as 'Space Gandalf'.

I have the greatest respect and admiration for science and the scientific community. I hero-worship the likes of Fred Watson and David Malin, and I'm amazed by the knowledge of the Professor Brian Coxes of the world to grapple with the endless complexity of particle physics.

However, mine is a much more practical role being able to show others what is right there in front of their faces. As for the theories? I'm much more inclined to wait and see. I feel the patience and the humour of the universe as it watches us little humans trying to figure it all out. I can almost hear the stars chuckling amongst themselves on my many quiet nights staring up at them from my swag.

When I went to university back in the early 1980s, my take was that university could teach you heaps of things, and if you remembered enough of those things they'd give you a tick, a piece of paper at the end of it, and away you'd go. That wasn't what I wanted, so I took off. I was looking for things that I wasn't finding at university which is why I still spend plenty of time in a bushman's swag.

Although I studied at the Quicke University of the Sky, many people do in fact study astronomy at a wide range of tertiary institutions around Australia. However, it seems that practical navigation and understanding of the night sky is not a part of the courses. I know this because I have astronomy students and graduates come to Broome to volunteer at Astro Tours and learn the sky. They know maths and physics and how to look at a screen; they know how to analyse the light from a star and make their own interpretations of what that light is, but some of them have never even looked through a telescope before.

After one of my early Astro Tours presentations in the mid-1990s, a lady came up and introduced herself.

'I'm an astrophysicist,' she said, shaking my hand.

'Oh wow,' I said. 'How did you go tonight?'

'Y'know,' she said, 'until tonight I've never thought about space and the stars like that before. So thank you.'

With some academics and intellectuals, the trickiest thing is to get them out of their heads. I'll show them where things are in the sky and they'll still complicate it by trying to fit it inside their head.

Often I'll tell them, 'Hey! Get it out of your head – put it in your feet instead! Stamp your feet on the Earth because that's our real planet. You don't have to make a model of it in your brain, because you can stand on it.

'Look at the Milky Way. That's the real thing. And if you look at the moon over there and look at where the Sun is, you can see the relationship between the Sun and the moon and the Earth.'

Most of them look at me a little strangely to start with, and then you can see the light go on as they switch on to what is right in front of their eyes.

These are real, tangible things but our education system often teaches us to disregard what's right in front of us in favour of what we can calculate and simulate and theorise about. I see it as my job to bring some focus back onto the glaringly obvious.

Chapter 15

Diamonds in the Sky

Human curiosity about the cosmos often begins from a very young age. After all, we sing 'Twinkle, Twinkle, Little Star' to toddlers.

'How I wonder what you are . . .'

I kept asking myself that very question as I beavered away cutting lines and fixing machines in the north-western deserts of Australia. Not only did I wonder exactly what they are, but as I reclined in my swag every night I'd ponder how far away they are, how big they are, how many of them are there, and how I fit into the picture of their existence.

Some of the answers were rattling around in the glove box of my Kenworth prime mover – or, at least, in the battered copy of the *Penguin Dictionary of Astronomy* I kept inside it. Today there is an endless trove of knowledge about the stars available in any number of books on modern astronomy. Even a quick Google will uncover countless online rabbit warrens of fascinating information to disappear down.

For this particular book on modern and practical astronomy, however, I've chosen to focus on a few of the interesting things about the stars that have captured my attention over the years.

Stars are suns

In 2002 I was riding my pushbike through a grove of peppermint trees on Rottnest Island, where I presented my star tours over four summers. I was en route to grab a post-surf snack at the local bakery when a big, solid bloke waved me down.

'Excuse me, mate,' he began as I rolled to a halt next to him in the shade of the trees. 'I just have to tell you something. I was out on your star show last night. I'm 46 years old. I'm a farmer. I've been around and done a few things and I've lived all my life outside.'

He paused to make sure he had my full attention, then continued. 'I haven't stopped thinking about what you told us last night about all the stars in the sky being suns. I've been up all night and outside for most of it, just looking at them, and it is completely blowing my mind. Do you really mean to tell me that every star I can see in the sky is a sun?'

I nodded.

'And that every one of them has got its own planets going around it too?' he pressed.

'Quite likely,' I replied.

'That's just blowing me away,' he repeated.

The big farmer looked intensely into my eyes as he shook my hand, before wandering off with a far-off look on his face. I'm still humbled by the moment of personal insight he shared with me that day. It's a reminder that some of the most profound things I share with people are the simplest things. He gave me the gift of knowing how important it is to share these things with everyone, even if it just turns out to be a reminder for some of them. Every star in the sky is a sun. Just as our Sun is a star.

Our current idea is that stars are huge balls of hydrogen that squeeze so tightly at their centres that chain-reaction nuclear fusion is generated. Maybe they're also talking to each other across the plasma filaments that connect them in the Herschel and Planck images from the European Space Agency.

Each of these hundreds of billions of suns potentially has its own family of planets. All of them are too far away for us to actually see their planets. Also, it would be like trying to see a grain of sand on the other side of Australia, with a powerful torch with a lens the size of a coconut shining in your eyes from the same direction.

Even with a powerful telescope, the light source of a star drowns out any possibility of seeing a tiny planet that might be alongside it. Some very clever astronomers are beginning to solve this problem and they've so far identified quite a few thousand stars that have their own planets, confirming what many of us have always suspected, including Giordano Bruno.

Many of the stars we can see are far grander than our Sun, while others are smaller, cooler, hotter, denser or lighter. To scan the star-filled sky from Earth is to gaze at billions of different worlds, each one of them unique. We live among them and they surround us on all sides. We see them at night like diamonds in the sky, but it's worth reminding yourself the stars are still there when the sky is blue, just temporarily upstaged by the cherished glow of our nearest star.

Far, far away

My high-school maths teacher would be pleased to know that trigonometry has enabled humans to measure the distance to many far-off suns. 'Parallax' measurements are calculated by 'solving' triangles. A triangle drawn from one side of the Earth's orbit to the other, and then out to a chosen star, can be solved if we know the length of one of the sides of the triangle and two of the angles. Knowing the Earth–Sun distance gives us the length of one side of the triangle.

In six months the Earth moves from one side of its orbit to the other and our point of view out to the star alters ever so slightly. The angle of this altered view can be measured to solve the triangle and determine the length of the other two sides. This gives us the distance to the star. Go to the top of the class, Quicke!

The accuracy of the parallax method relies on our ability to measure very small angles and our ability to accurately gauge the Earth–Sun distance. For a long time parallax was only useful out to about 100 light years away, after which the angles became too minuscule for us to measure.

The smudginess of Earth's atmosphere also limited our fine measuring capability, so in 1989 the European Space Agency launched the High Precision Parallax Collecting Satellite, or Hipparcos. This increased the accuracy of parallax measurements to around 1600 light years! Since then, the ESA's Gaia mission has extended this accuracy across the 100,000 light years of our galaxy and beyond to our local group of galaxies.

Prior to Hipparcos and Gaia, the estimated distance to Canopus – the mighty elephant-sized star that is halfway to the moon in our sand-grape-coconut model of the solar system – varied between 100 and 1600 light years. Hipparcos gave us a 3D position among the stars and a far more accurate impression of some of our next-door neighbours. Canopus, we now know, is 310 light years away.

Stars shine bright in our sky either because they're close to Earth or because they're big. Or humungous in some cases. Whether they're near or far, one way we categorise stars is by their brightness in our sky. First-magnitude stars are twice as bright as second-magnitude stars, which are twice as bright as third-magnitude stars and so on.

If you're lucky enough to have excellent eyesight, the limit of visibility under perfect, dark sky observing conditions is the faint stars at magnitude six. A pair of 50 mm binoculars will fill the sky with even paler stars down to magnitude 11, while a 10-inch telescope is capable of filling in a few more of the gaps with stars that are dimmed way down at magnitude 14.

We have 21 first-magnitude stars in our sky. Four of these shine strongly enough to be measured with *minus* magnitudes. At magnitude –1.44 Sirius is the brightest star we can see from Earth. By contrast Venus can be the brightest planet visible when it reaches –4.6 at the right time in both of our orbits. Jupiter comes in at –2.8 when it is on the same side of the Sun as we are. The full moon can reach –12.6 and the Sun is a consistent magnitude –26.73. All of the visible planets except Saturn are capable of achieving negative magnitudes, depending on their relationship to us and the Sun at the time.

For now, though, let me shine a light on a few of these celestial wonders 'up above the world so high' in order of brightness, so you can get to know them a little better. Where I can, I'll tell you where to find them in the sky.

Sirius: magnitude –1.44

Twinkling the brightest is Sirius, which dominates our sky partly because it's relatively close at 8.6 light years away and partly because it's twice the size of our Sun – and a whopping 25 times brighter.

There are only six stars closer to us than Sirius, although with the discovery of more dim, 'brown' dwarf stars, we're likely to find a few more as technology and knowledge progresses. All of the closest six, with the exception of Alpha Centauri, are too dim to see with the naked eye. This may sound counterintuitive, but it's a very common story, considering 85% of the stars in the universe are thought to be smaller and fainter than our Sun. We simply don't see them at all.

Any stars further than 55 light years from us must be bigger than our Sun for us to see them, which means our Sun would be too dim to see from 55 light years away. Many of the stars in our sky are much further away than 55 light years, which means most of the stars we see in our sky are very, very big and a great deal mightier than our Sun.

Like our nearest neighbour Alpha Centauri, Sirius is a double star. Sirus A is by far the larger of the two. Sirius B is a little white 'dwarf star' the same size as Earth, and the same mass as our Sun but 50 times fainter. A handful of Sirius B substance would weigh many tonnes. You won't see Sirius B without some seriously dedicated telescope work.

Sirius A and B are about 2.5 light hours apart and go around each other once every 50 years.

At 17 degrees south of the Celestial Equator, Sirius can be seen from most places on Earth. You'd have to be well above the Arctic Circle before Sirius failed to show. To find it, follow south from a line drawn through the familiar three belt stars of Orion the Hunter and you will come pretty much straight to Sirius. It's the brightest and most obvious star there.

Canopus: magnitude –0.74

Canopus, the second brightest star in our sky, is an amazing 310 light years away, yet it's brighter than Alpha Centauri at only 4.3 light years away. Canopus is by far the brightest star within 700 light years of Earth. Sirius only appears brighter to us because it is much closer at 8.6 light years.

Shining 15,000 times brighter than our Sun and being 65 times bigger, Canopus would occupy the space almost all the way to Mercury if it was in the same place as our Sun. For Canopus to appear the same size as our Sun does to us, we would need to be three times further away from it than Pluto is from our Sun.

Canopus is deep in the southern hemisphere at 53 degrees south of the equator, so you'll need to be somewhere south of the Tropic of Cancer to get a decent look at it. From the tropics region in the northern hemisphere, Canopus will graze your southern horizon in the early evenings early in the year. Northern hemisphere navigators can use Canopus as a type of southern 'pole star', because when it's at its highest altitude it indicates due south.

Along with Achernar, the Southern Cross and the Pointers, Canopus is very much a star dominating the south polar region. Follow the same line through Orion's Belt to find Sirius, keep the line going – bending it a little to the left – and Canopus is again the most obvious bright star in that direction.

Alpha Centauri: magnitude –0.28

Alpha Centauri is about the same brightness as our Sun. If we were to go there, our Sun would look about the same to us as Alpha Centauri appears from here, yet we would hardly have gone far enough to alter our perspective on the rest of the stars in the sky. Even Sirius would look much the same from Alpha Centauri.

If we could travel 55 light years, both Alpha Centauri and our own Sun would be invisible to the naked eye. If you are lucky enough to live in the southern hemisphere, Alpha Centauri is easy to find. It's the brighter of the two pointer stars associated with the Southern Cross. Alpha Centauri looks like any other star in the sky until you point a telescope at it. Any small telescope will show you two stars, very close together.

Arcturus: magnitude –0.05

The fourth brightest star in our sky is one of my favourites. Arcturus is a good example of a maverick star that wanders the galaxy on its own. Fixed stars won't appear to move in relation to each other in a human lifetime, despite the fact they *are* on the move. Some of them travel in groups and some go it alone. We call the movement of stars 'proper motion'.

Around half a million years ago, Arcturus was too far away to be seen from Earth. It's currently wandering past us at a distance of 37 light years, but in another 500,000 years it will have drifted out of sight again. For the rest of your life, though, you'll find Arcturus in the constellation of Boötes, 19 degrees north of the equator. It's the brightest star in the northern hemisphere and visible to almost everyone on Earth.

Vega: magnitude 0.03

Like Arcturus, Vega is also in our local neighbourhood at 26 light years away. Since it's 39 degrees north of the equator, most people in the world can see it. If you live in Tasmania you'll need a clear northern horizon in July, August and September, and look for it in the evening sky.

Vega was one of the first stars around which astronomers detected a disc of dust, which suggested it has a solar system similar in structure to our own. We've since discovered many other dusty discs and planets orbiting other stars.

Capella: magnitude 0.08

Capella is 42 light years away from Earth. Since it's 46 degrees north of the equator, seeing it from southern Australia requires some planning. Capella is a summer star for southern observers so a clear northern horizon will show it well in January.

Rigel: magnitude 0.15

When we jump way out to the seventh brightest star 860 light years away, we find the first sun on our list to rival Canopus in size. At 78 times the size of our Sun and 40,000 times brighter, Rigel is even larger than Canopus. Everyone in the world can see Rigel as it's only 8 degrees south of the equator in the well-known constellation of Orion the Hunter.

Procyon: magnitude 0.38

Procyon is 5 degrees north of the equator and really quite close by at only 11 light years away. It's the brightest star in the constellation of Canis Minor, or the 'little dog'. Sirius and Procyon are the two hunting dogs of Orion straddling either side of the Milky Way.

Achernar: magnitude 0.45

Achernar, the ninth brightest star, is 144 light years away, and, along with Canopus, it very much dominates the south polar region. Achernar is opposite the south celestial pole from the second of the pointer stars, Beta Centauri, and almost the same distance from the pole itself. Eridanus the River is a long winding constellation that begins at Achernar and ends way across the sky near Orion.

Betelgeuse: magnitude 0.5

Coming in at number 10 on the brightness list, Betelgeuse fires the imagination as much for its funky name as for its incredible size. More than 1000 times bigger than our Sun, this star would swallow Mercury, Venus, Earth and Mars if it was in the same place as the Sun.

Betelgeuse is a 'red supergiant star' 640 light years away. The familiar three belt stars of Orion are flanked on either side by two first-magnitude stars. Betelgeuse is distinctly red, and the other, Rigel, is so white it's almost blue.

Beta Centauri: magnitude 0.61

Two stars that help us get in touch with the dimensions and depth of the cosmos are the pointer stars. They look very similar out there, pointing neatly towards the Southern Cross. The difference is that Alpha Centauri is 4.3 light years away while Beta Centauri is 390 light years away! Does that make them look different to you?

Beta Centauri is the 11th brightest star in our sky and one of only a small handful of bright stars anywhere near the south celestial pole. It is the closest pointer star to the Southern Cross.

Special mention: Fomalhaut: magnitude 1.15

Fomalhaut is a first-magnitude star only 25 light years away in the constellation of Piscis Austrinus. It comes in at 18 on our list and merits special mention because in May 2008 images from the Hubble Space Telescope showed a planet in orbit around it – the first time another planet was imaged near another star in visible light.

By now we know with some certainty that all of the suns in our skies are solar systems, each with their own families of planets. Efforts to find them in the last decade or so have met with spectacular success. Almost all of these discoveries have been with indirect observations, so having our first visual confirmation was exciting indeed.

Where do we fit in?

Of the 10 brightest stars in our sky, six are within 50 light years. The other four range out to nearly 1000 light years. There are

20 stars within 12 light years of Earth and the 30 brightest stars in our sky range way out to Deneb at 3229 light years away.

While there are 21 first-magnitude stars, there are many more second-magnitude stars, and around 6000 stars are visible to the naked eye under perfectly dark conditions. At any one moment, though, you, me and the burly farmer I met on Rottnest Island will only see around 3000 stars, because the rest will be below the horizon.

The next time you're out gazing at them, take a moment to look at your own body and realise all of the elements that make up a human being can be found in the billions of suns around us. We are, quite truthfully and simply, made of stardust.

Chapter 16

Where a Star Is Born

Not only are we made from the same stuff as stars, we follow the same pathway of birth, life and death. And just as humans can be categorised according to gender, age, race, nationality, size, attitude and whether we bounce out of bed in the morning or need three cups of coffee to get moving, so too can the suns of the universe be pigeonholed.

Our Sun falls into the category of 'dwarf stars', although being brighter than 85% of other suns it is at the higher end of this cohort. Most of the stars we see in our skies are in the top 15%, simply because they're bright enough to be seen.

There are myriad categories including brown dwarf stars, white dwarf stars, blue giant stars, red giant stars, pulsars, neutron stars, black holes, double stars, triple stars, and almost any variety you care to imagine. Astronomers have even found a 'diamond star' – made of crystallised carbon like the gem here on Earth – and I reckon it's only a matter of time before they find a star made of marshmallow.

The bigger stars in the universe tend to live fast and die young, blowing themselves to smithereens in as little as 10 to 20 million years. Other stars pursue long, relatively boring lives and exhibit predictable, plodding behaviour. Fortunately, our Sun is a plodder. This has allowed humans ample time to get on with quite large projects like, you know, civilisation! Life on planets orbiting more volatile stars would be way more challenging and involve being barbecued one minute and frozen solid the next.

Our Sun is about five billion years old and it's estimated to be middle aged. The good news is this gives us another five billion years to finish all those projects we've started.

Nebulae – star nurseries

Stars are born inside enormous clouds of ionised gas or plasma and dust called nebulae. The material is mainly hydrogen, with a little bit of helium and lithium and a smattering of all of the elements of the universe, depending on whether they are second or third-generation nebulae.

In one sense nebulae are the rubbish dumps of space; vast holding grounds for material left over from previous solar explosions and galactic disasters that occur in the natural course of universal history. A star is born when a pocket of dust and gas within a nebula condenses. This condensing activity could be triggered by an external event, like a nearby supernova explosion that sends a shock wave through the cloud, and causes some of the hydrogen atoms to bump into one another. The magnetic filamentary streamers shown to us in beautiful detail by ESA's Herschel and Planck missions and the stars strung along these filaments within star-birthing nebulae certainly provide insights into how this might be happening.

The theory of accretion suggests that this kind of bumping interaction causes two atoms to stick together. Bring in the ideas of electromagnetism and we have attraction and repulsion at work, rather than only the one-directional gravitation force. Gravity attracts and does not repel. Electromagnetism both attracts and repels depending upon its charge, and so gives it the ability to organise material into layers and filaments, as we see in plasma laboratories on Earth and now from Herschel and Planck.

Accretion theory also suggests that these new buddy atoms weigh twice as much as the surrounding atoms and thus have twice the gravity. This attracts a third atom and then 10 more, then a few hundred, and before you know it you've got a big ball of atoms stuck together and dominating their surrounds with gravity.

With the electromagnetic forces in an atom being 10 to the 36th power – that is, 10 with 36 zeros – times stronger than the gravitational forces, their participation in organising material into this star-birthing process seems likely.

The theory goes on to say that new atoms are attracted from places further afield and, with more room to pick up speed as they're pulled in, they slam into the ball so fast and from so many directions they cause the whole thing to spin. Every new atom contributes to the spin, mass and gravity of the ball. The rotation accelerates to the point that some groups are flung from the main mass to swing around the outside on their own. We live on one of these bits.

While the central mass of the embryonic star grows, the pressures inside steadily increase as more and more material joins the big squeeze. Eventually two of the atoms somewhere deep in the centre of the mass of hydrogen are pressed so close together they fuse to form an atom of helium. The energy of this nuclear fusion sets off everything else around them. Before you know it, there's a chain-reaction nuclear fusion and, *voila*, a star is born!

We've already shown that this campfire nuclear furnace in the middle of a star model doesn't give us a clear picture of what's going on. Combine it with the relatively new discoveries of magnetism and electricity, with their ability to organise the plasma that permeates all of space, and maybe accretion begins to make more sense.

What we do see inside nebulae is that the growing solar wind from a new star gradually blasts a hole in the nebula, making room for itself to shine across space where we get to enjoy its twinkle on a clear night from Earth.

Nebulae are constantly changing environments, shifting and morphing as the material within them is spent in the creation of stars. Eventually all the dust and gas is used up, leaving just the stars born from it. At that point the star nursery becomes a star cluster, with any remnants of nebulosity blasted away into space by the solar wind from the newly forged suns.

The Orion Nebula

One of the closer nebulae to Earth is the Great Nebula in Orion. You can see it with the naked eye on a clear, dark night as the fuzzy middle star in the Sword of Orion. A pair of binoculars will show it even more beautifully and it is a favourite object for telescope observers all over the world. At 1500 light years away and 24 light years across, the dust and plasma cloud of the Orion Nebula is an enormous star-birthing region.

The Hubble Space 'Telescope in particular shows us deep inside this plume of dust and ionised gas to reveal little pockets blasted in the cloud by new stars.

The Trapezium Cluster is a delightful little diamond-shaped group of four stars in front of the Orion Nebula. Over the last few million years these young, fierce stars have blasted a big hole in the nebula, clearing the way for us to see them at all. In return, they shine their light on the plasma cloud of the nebula itself, reflecting that light off the cloud and shining it all the way across space, so we can see it from here on Earth.

The dark, the light and the colourful

Nebulae reveal themselves in three different ways. We can see the Orion Nebula because the stars shine on it, which in turn reflects the light back to us. This is a reflection nebula.

Emission nebulae glow in the same way a fluorescent light tube glows when electricity flows through it. Hot, energetic stars nearby the cloud of gas and dust bristle with electricity and strip electrons from the molecules of the nebula, causing them to be 'excited' to the plasma glow state. Different gases in the nebula emit different colours. Oxygen shows up as green and hydrogen as red.

Dark nebulae are exactly the same loose and irregular clouds of gas and dust as the reflection and the emission nebulae, yet they are not lit up by any nearby stars. Instead, they are conspicuous by their apparent absence and reveal themselves as black patches against the bright star fields of the Milky Way.

The Coal Sack

There appears to be a dark hole in the Milky Way, right next to the Southern Cross. It is, in fact, the Coal Sack – a huge interstellar cloud of dust and gas that's 500 light years away and big enough to block out the light from the Milky Way. The Coal Sack appears as big as the Southern Cross, and while it isn't the largest nebula around, at 60 to 70 light years across it is close enough to fill more of our sky than any other nebulae.

Marala, the emu man

The Coal Sack is also the head of the dark emu in the sky, known in my part of the world as Marala, the emu man. According to local aboriginal lore, Marala made everything on the Earth and taught the people right from wrong, before flying up into the sky where we see him today.

His head is the Coal Sack while his neck is a large dark streak through the constellations of Norma and Circinus. His back and his belly balloon out, again as dark parts of the Milky Way, around the Scorpio/Sagittarius region, before his legs split on their way down towards the constellation of Aquila the eagle.

Once you've seen the emu stretching nearly 45 degrees across a dark sky, you'll wonder how you hadn't seen it before. Just remember, while you are looking at him, he is also looking at you, to make sure you know right from wrong.

Planetary nebulae and supernova remnants

These add to the mix in this most intriguing group of some of the biggest structures in the galaxies. The first three types of nebulae are very similar in make-up, structure and purpose. They are plasma clouds and they are star nurseries. The other two types – planetary nebulae and supernova remnants – are generally smaller, fainter and harder to see.

Planetary nebulae seem to be stars that are ending their lives by blasting out shells of glowing ionised material. They display many varied patterns as revealed by a series of images from the Hubble Space Telescope. We see some of them as discs of light, which is why they were saddled with the 'planetary' name.

Supernova remnants are the result of stars blowing themselves to bits even more violently and scattering their material far and wide across the galaxy. Eventually this discarded matter finds its way into the star-birthing rubbish dumps – the nebulae – to be recycled back into the next batch of solar lives.

What a system!

Star clusters – local villages

After all of the gas and dust of a nebula is used, we're left with a star cluster. When they're young these new suns tend to stick together as compact family units, just as we do here on Earth. As the cluster matures, the stars spread out, just as we do. After a while, some of the stars leave home and wander the galaxy on their own, again just like us.

Many of these young, compact, galactic open star clusters, and their older cousins, the more diffuse open star clusters, fill our night skies. You can easily find them with a casual search with a common pair of binoculars. Once you've found them that way, you'll know where to point your telescope for a closer look.

Hyades cluster

The closest of the open galactic star clusters to us are both found in the constellation of Taurus. The Hyades star cluster, which forms the head of Taurus the bull, is a neat triangle of around 130 stars with the 12th brightest star in our sky, Aldebaran, dominating.

Aldebaran – known as the 'bull's eye' – is an orange giant star that is not actually part of the Hyades cluster. It is situated well and truly in the foreground at 42 light years away, while the magnificent star field of Hyades itself is more like 150 light years away. That's pretty close! The Hyades stars are easily visible with the naked eye and come up beautifully in your binoculars.

Pleiades cluster – the Seven Sisters

At 400 light years away, the Pleiades is our next nearest cluster. They, too, are found in Taurus and are also known as the Seven Sisters from not only Greek mythology, but in almost every indigenous culture on Earth. The Pleiades are easily seen with a casual glance around the southern summer sky or the northern winter sky. On a moonless night you can hardly miss them.

Our direct vision is less light sensitive than our peripheral vision, so the Pleiades – which are mainly fourth-magnitude stars – are at the lower end of our direct vision's capability. Glance to the side of them, however, and they will be clearly visible. Snap your vision back onto them and they will fade out and disappear.

Peripheral vision can be used to advantage when looking at faint objects, by simply flicking your eyes back and forth across whatever you're looking at. It's a great technique when using a telescope on faint objects too.

Most people with average eyesight can see between six and nine stars in the Pleiades. A pair of binoculars will reveal at least 100. The Pleiades and the Hyades are two of the most beautiful sights in the sky. Both are close enough to the equator that everyone in the world can see them.

A galaxy of star clusters

A pair of binoculars will reveal many other open star clusters scattered all over the sky. The beehive cluster in Cancer is only 500 light years away, the Jewel Box near the Southern Cross is 8000 light years away, and there are many, many others in between. Telescopes bring them to life and show an incredible variety of shapes, sizes, brightness and colours.

Whenever you look at star clusters, bear in mind that every one of the stars you can see – even the tiny ones in the background – is an entire solar system!

Globular clusters – stellar cities

Some things are stranger than science fiction. Where open star clusters are loose groups of a few hundred stars at most, the globular clusters are compact balls of hundreds of thousands of stars, sometimes millions.

Stars in globular clusters tend to be very old, with some calculations putting their origins before the so-called 'Big Bang'. Problem? Indeed! Instead of arriving via the regular star-birthing activities that occur in nebulae, the 200 or so globular clusters in our galaxy seem to have sprung into existence all at once and very early in the life of our galaxy.

Omega Centauri

The biggest and most impressive globular cluster we can see from Earth, this mega-city of stars is clearly visible with the naked eye deep in our southern skies in the constellation of Centaurus. The first time I found the fuzzy ball of light that is Omega Centauri in a pair of binoculars, I knew it was time for me to get a telescope. I desperately wanted to get close enough to see the individual stars in the centre of this most amazing of the globular clusters.

Current thought and research is that Omega Centauri is a ball of 10 million stars, 200 light years across and 16,000 light years away. Who counted them all? There are some people around who haven't discovered surfing yet and have more time on their hands than me.

47 Tucanae

The next best globular cluster is also found deep in southern skies. With only one million stars, 47 Tucanae rivals Omega Centauri for sheer splendour, due to a very compact core of stars surrounded by a looser spray of suns in its outer reaches. By contrast, Omega Centauri is big and uniformly populated right across its width.

The best of the northern hemisphere globular clusters is in Hercules and it pales into insignificance compared to these two giants of the south. Both can be seen with the naked eye and binoculars bring them up beautifully. Telescopic views are unbelievable; you can dive deep into the centres of these celestial mega-cities to check out individual stars.

Chapter 17

When Stars Die

One day our Sun will die and, along with it, so will the Earth as we know it. Maybe the stars eventually run out of the material they turn into energy. Or they run out of squeeze to meet the ever-increasing demands for greater levels of nuclear fusion. Or the fuel load they carry is greater than their gravity can contain. Or they are simply cut off from whatever is powering them.

Whichever way it happens, the end result is death. But just as our ashes are to ashes and our dust is to dust, the death of stars ultimately leads to astral recycling as one crop of dead suns provides the material for the next generation of bright, burning plasma balls in our amazing, ever-evolving universe.

Some stars shake a metaphorical fist at death, throwing hissy fits and casting big portions of themselves out as shells of material. As this expanding matter cools, the star appears to redden. This can occur a number of times before the star gives up the ghost entirely, resulting in some truly breathtaking shapes and patterns in space. Planetary nebulae, we are looking at you.

You'll need a telescope to see any of the planetary nebulae and activating your peripheral vision for these faint, fuzzy smudges will pay off too. The Hubble Space Telescope amassed a gallery full of beautiful and colourful images of planetary nebulae.

Supernovae

Other stars have too much gravity for this expansion to be successful, and they collapse in on themselves so violently that they explode as supernovae.

The incredibly violent and cataclysmic nuclear processes arrived at in a supernova form heavy metals that otherwise couldn't be formed in the usual processes going on inside stars. According to the theory, most stars can build all of the elements up to, and including, iron. However, once iron starts to form in the core of a star, it holds back any more advanced nuclear fusion. There isn't enough gravity to squeeze the iron atoms together to fuse them into the next element.

Many stars whimper and die at this stage, fading away to become dark cinders wandering aimlessly through space. Some of them will cast off rings of material as planetary nebulae. In supernovae, however, we think that gravity wins the battle. Expansion happens for a while, then the gravity takes over and accelerates all of the expanded material back into the centre of the star, where the rebound sets off an explosion with spectacular results.

The creators and annihilators

A supernova explosion is so violent that advanced heavy metals are fused or created. Iron is released and fused into heavier elements. The violence is so great that the star blows itself to bits, scattering its remains – and all that it has created – far and wide through its vast local region of space.

It's been over 400 years since we had a supernova go off anywhere near us. Kepler's Supernova of 1604 was 20,000 light years away and yet was brighter than every star in our sky, with only Venus outshining it. It was still visible to the naked eye 18 months later.

Chinese observers recorded a supernova in 1054 that was 6500 light years away in the constellation of Taurus. The supernova SN 1054 left us a remnant that we can still see as the Crab Nebula. As you can see, two supernovae in the last thousand years makes them a rare event in our galaxy.

Most supernovae we see are in other galaxies, where they outshine their entire galaxy of billions of stars for a few weeks as they let go. The closest one in recent times was SN 1987A, which went off in the Large Magellanic Cloud galaxy in 1987. Actually, to get here by 1987, the light from this major galactic event had already been travelling for 170,000 years!

Are you ready for the next supernova?

The next supernova in our own galaxy may come from one of our neighbouring red giant stars, Antares in Scorpio or Betelgeuse in Orion, on opposite sides of the sky. We're watching them quite carefully as at least one of them is likely to go off sometime soon. By soon, we mean sometime in the next 10 to 20 million years. They might go off tomorrow too!

At 600 and 860-odd light years away, both of these stars are much closer than any of the supernovae in our recent thousand-year history. They're both such big stars that if either of them were in the place of our Sun, they'd stretch out all the way past Mars and halfway to Jupiter.

At the time of writing in 2020, Betelgeuse has recently given us cause for concern. Speculation erupted in October 2019 about whether Betelgeuse was about to go supernova, as it faded to 40% of its usual brightness as the 10th most luminous star in our sky. By May 2020 it had started to resume normal transmission, and unless it has gone off between now and publication of this book, then all is well in the world that is Betelgeuse.

If gravity eventually wins the battle with the expansion processes happening inside these giant stars, a supernova is likely to result with spectacular effects from our grandstand view here on Earth. When they do go off, they will probably be the brightest objects in the night sky for two or three weeks – so bright they will be easily seen during the day.

If they go off while we're still here watching, it will be very interesting indeed. We haven't had anything this energetic happen this close to us in our known history. We are likely to be bombarded with photons, x-rays, gamma rays, cosmic rays and other energies in amounts and at levels that we might never have dealt with before. Current scientific thought is that we will be okay with this, and that containing our excitement at the magnificent sight of giant exploding stars will be our greatest challenge.

Ringside seats, anyone? Anywhere on Earth will be perfect.

Atoms of iron – the truck drivers of life

We rely on supernovae to provide us with the heavier and more complex elements of existence. In scattering their remains in their cataclysmic explosions, they provide the material for the next generations of stars. Our own solar system is one of these next generation stars. We're all made from ex-supernova material.

We can be sure of this because of the abundance of metals on this little planet of ours. All of the iron on Earth was made in a supernova explosion. We haven't figured out any other way of making iron. Have a look at your car, have a look at the next bridge you drive over, have a look at the metal that is all around you every day.

There's iron in your body too. Haemoglobin molecules are coursing through your veins right now. Each one of these molecules has an atom of iron at its centre. Haemoglobin is what makes your blood red. There's an electrically ionised atom of iron at the centre of every haemoglobin molecule that attracts and bonds with the oxygen molecules in your lungs. This is going on right now, while you read these words.

Having picked up its little parcel of life-giving oxygen, the iron atom carries it quite happily until it encounters a cell in your body with a different electrical charge and they make an exchange. The oxygen tips through the cell wall to do its job of re-energising the amazing little worlds that your cells are, and the waste carbon dioxide now has the right charge to be picked up by the iron. The iron carries the rubbish back to your lungs and you breathe it out!

Every one of those atoms of iron, rushing around in your

body like little dump trucks, originated in a supernova explosion. How are you feeling? Take a few deep breaths and charge those iron atoms with oxygen right now.

Recycling is out there already!

Supernova material eventually gathers in the vast rubbish dumps that we see in the sky as the nebulae that go on to produce the next generation of stars. These rubbish dumps are still mostly hydrogen and helium, although, being second-generation solar material, contain a peppering of heavier elements too.

Our own solar system is at least second generation, with all of its matter having been through another solar system before, if not being a mixture of a number of different solar systems. It's all about recycling. Even the body you inhabit is made of recycled materials that have served in at least one other solar system before this one.

Black holes – getting degenerate

Supernovae might even leave behind a small part of themselves, which goes on to form a black hole or a neutron star. According to black hole theory, the gravitational pressure in these remnants is so great that the regular atomic material, as we know it, is degenerate. The vast spaces usually present in between protons, neutrons and electrons in an atom of matter are reduced, until all of these subatomic particles are bang up against each other. The mass that would be spread out over the vast distances of emptiness inside regular atomic material – the material that we're used to dealing with in our daily lives – is now compacted to the point where a teaspoon of neutron star material would weigh thousands of tonnes!

A black hole takes this to even greater extremes, with the material occupying a point of infinite smallness and exerting infinite gravity in this infinitely small space. Make sense? No, of course it doesn't. The universe is still full of mysteries and black holes fit really well into this category. I've heard it suggested that they're simply doorways or wormholes into other universes. With a planet-sized diamond out there, anything is possible!

You won't see a black hole or a neutron star with a casual glance around the sky because, well, they're black. We've found places that are best explained by the presence of a black hole, because things are going around a seemingly empty place very fast. Things can go into orbit around a black hole in the same way they go into orbit around our Sun. They can mess up and crash into the Sun too, and many comets do just this. Anything that comes too close to a black hole crashes in and doesn't come out either.

Escape velocity from Earth is 11.2 km per second. This is the speed you'd need to blast a rocket off from our planet to escape its gravity. If you go slower, your rocket won't make it. It will turn around and come crashing back to Earth. Black holes are black because even light, travelling at 300,000 km per second, isn't fast enough to escape. The escape velocity from a black hole is faster than the speed of light.

If you shine a powerful spotlight up from inside a black hole, it will fail to achieve escape velocity and your light beam will turn around and crash back down into it. An exception is the supraluminal (faster than the speed of light) jets of plasma that seem to be shooting out of the poles of some black holes. Are you getting the idea that there are a lot of things we don't know yet? We've got a lot to learn. Me? I'm happy to wait and see.

Black holes are weird.

A galactic menagerie

These are just a few of the things that make up the population of a galaxy. Some of them are pretty weird and I'm sure there are even weirder things to find.

We're becoming familiar with the stars as solar systems with their families of planets, comets, asteroids, rocks, dust and plasma. We're familiar with at least one of these planets, with its dogs, cats, volcanoes, Harley-Davidsons and giraffes, with its mountains, oceans and rainforests, with its cities and ice caps and aeroplanes and whales. We live here. We know about other planets because we can see them in the sky if we know where to look.

The stars, including our own Sun and its solar system, belong to a family group of other solar systems forming star clusters. The globular star clusters are the mega-cities. Nebulae are both the rubbish dumps and the birthing suites of existence. Just like each one of us, each star is a unique individual. Just like us they are born, they live their lives and they die. Just like us they do this in an amazing variety of ways.

Put all of these things together and we have a galaxy, once called an island universe, isolated from the other island universes by vast areas of seemingly empty space, intergalactic space, or the space in between the galaxies.

Is it empty space? Probably not, although it is empty and transparent enough that we can see all the way through it to the other galaxies scattered in every direction in their teeming millions. Imagine being on a planet going around a rogue go-it-alone star out there in intergalactic space. If we were halfway between the Milky Way and Andromeda, for instance, there would hardly be any stars in the sky at all and these two giant galaxies would only be faint smudges.

I, for one, am grateful to be embedded deep inside a populated region of the Milky Way galaxy, with its unlimited entertainment value in showing us such a variety of other worlds. Having a planetary atmosphere clear enough to see through, so we can have the knowledge of the other worlds, is surely a heaven-sent bonus for anyone who has ever looked up in the sky and wondered.

Chapter 18

How to Watch the World Turn

The most fundamental thing I share with audiences at Astro Tours is that they are on a planet that is moving – *really* moving. Whenever I point this out, however, people invariably respond, 'Why can't we see it?' I assure them that not only *can* they see it, they already *do*. Every day.

Whenever you see the shadows lengthen or the Sun appearing to slide slowly across the sky, you are looking at real-time evidence of the planet spinning around on its axis once a day. You can tune in to this at night by watching the moon drift from horizon to horizon.

On nights when the moon isn't visible, the stars are always there to silently and methodically chart our movement. All you have to do is pick out a few prominent stars, give them whatever names you like (although a star chart will soon have you reciting their actual names like a pro), and watch them for half an hour.

If it's dark in your part of the world right now, why not duck outside and select a couple of stars this minute? If it's not yet night-time, maybe you could come back to this chapter when it *is* dark.

Okay, once you've selected your stars line them up with a tree, a mountain or the side of your house, so you can easily locate them again. Now, go and make a cup of tea or whatever, wait half an hour and stand in the same spot to check the position of your stars again. They will appear to have moved. They haven't, of course, but the Earth and all who sail on her have.

So there you have it! You've just watched the planet spin on its axis over a quick cup of tea.

Fixed stars

How do we know it's us and not the stars that are moving? The stars are staying still enough in relation to each other that you won't notice any change in those relationships for the rest of your life. You can rely on them. Start watching now and see if I'm right. We even call them 'fixed stars'.

The deeper truth is that the stars *are* actually moving in relation to each other, but they're so far away you won't notice any change in just one lifetime. Were you to live 400 or 500 lifetimes, you might be able to chart this glacially slow stellar migration, or what's known as the stars' 'proper motion'. As I'll come to later, some stars will journey through space together, while others will wander off on their own path. I guess they're a bit like us in that regard. Or maybe we're a bit like them. At any rate, during our fleeting human lifetimes, the stars remain as good as perfectly still. That means we can use them to watch our own movement amongst them.

The wanderers

The stars may be fixed but the planets are constantly and visibly on the go. Indeed, the word planet derives from the Greek *planete*, which means 'wanderer'. Earth is one of these great travellers too.

Five of the planets in our solar system are bright enough for us to see from Earth, and are easy to follow as they track through the fixed stars. This isn't something you'll notice in one night of casual observation; the planets aren't moving quite that fast. But if you watch them from one night to the next, you'll track their movement with ease as they orbit the Sun, just as we do.

It's particularly easy to follow the faster-moving planets Mercury, Venus and Mars. You'll be onto them in only a few nights of casual observing. No need for a fancy telescope or even binoculars; all you need are your eyes and for those planets to be above the horizon when it's dark enough to see them. If you need help locating Venus, Mercury or Mars, grab a star chart or an app for your phone. Then, simply line them up with some other bright star in the sky and see if they've changed position by the next night. (Hint: they most certainly will have.)

Jupiter and Saturn – the other two planets we can see with our own eyes – dawdle much more slowly through space, so they appear to be almost fixed with the stars for weeks or months at a time. But they *are* moving – you just need to be patient in your observations.

Our heading through space

To tune in to Earth's celestial trajectory, it's useful to know the directions of north, south, east and west. That's easy if you have a compass, but if you don't, it's not too hard to work out for yourself. The points of the compass have some fairly basic and reliable markers. It might sound silly but it's worth stating that the Sun appears in the east in the morning and disappears in the west in the evening. Most maps have an arrow that points north. All you have to do is line the map up with the walls in your house or the direction of your street, to take note of where north is. To find south, simply about-face 180 degrees.

Now that you're a finely calibrated human compass, turn towards the east. South is to your right, north is to your left and behind you is west. Since you're now looking east, you're facing in the direction of the Earth's daily spin cycle through time and space. You're going forwards at over 1000 km/h. Can you feel the interplanetary wind in your face?

At an arm's length

Your hand, held out at arm's length, makes a terrific astronomical measuring stick. With your arm extended straight out, stack your fingers on top of each other or spread your thumb and fingers. Close one eye and line your hand up between the horizon and any star you can see in that eastern sky.

Each finger is about 2 degrees wide at arm's length. From your thumb to your stretched index finger is about 20 degrees. It's more like 25 degrees if you use your thumb and little finger.

The numbers aren't really as important as simply using your hand as a way of measuring how far apart things appear to be in the sky. You can use the same method to watch planets moving in relation to each other and in relation to the stars too.

Let's do the same in the west. Pick any bright star and measure its height above the horizon. Since west is the direction that we're turning away from, the stars will appear to be heading for that horizon.

The next direction to consider is straight up. Pick a bright star somewhere up high in the sky. The best way to measure this is to see how much your neck hurts. You can use the moon for any of these measuring exercises if it happens to be around.

By watching carefully and using your fingers as a ruler, you will easily notice movement among all of your selections in just 20 or 30 minutes.

Looking out vs. looking up

When we look up from the surface of the Earth on a clear night, we're in fact peering out through a wonderful window all the way across interplanetary, interstellar and intergalactic space. It's a breathtaking vista that extends for trillions of kilometres.

I say we peer 'out' at space because, more often than not, we call the direction over our heads 'up'. The tricky thing about that is in 12 hours' time, after the Earth has turned halfway around from where it is now, the direction you just called 'up' will be 'down'.

Take a minute – while the Earth turns you a bit further through space – to think about where you were 12 hours ago, where you'll be in another 12 hours and where the Earth is taking you in every moment of your life.

In astronomy it's also useful to get your head around this direction of 'out', because we're on a moving platform. 'Up' and 'down' are forever changing, and as the Earth endlessly turns us around, we're offered a constantly shifting view out into space.

Now you've got three of the key directions sorted out: east, west and straight out (up).

Spinning right out

Because Earth is turning from west to the east, the stars – which are fixed locations in the sky – will appear to move from east to west. There are, however, two more directions to explore and this is where things start to get interesting.

The Earth spins around its north and south poles. This is the Earth's axis. If you turn yourself to face south, the Earth will be turning you to the left and the fixed stars will appear to turn to the right. Face north and the Earth will turn you to the right and the stars will all appear to turn left. Stand facing east or west and extend your arms to point north and south, and you're beginning your orientation with the Earth's axis of rotation. Congratulations!

What I've just shown you, though, with your arms pointed at the northern and the southern horizons would only be completely true if we were standing at the equator. As soon as we do that little thing that's called 'travel', all of this changes. If you travel around the curve of the Earth to go south, the stars will all appear to move north. If you travel north, they'll appear to move south.

North and south

Travelling north and south is just like walking over a hill and being able to see what's on the other side. You go further around the curve of the Earth – the metaphoric hill – and see the stars that were below the horizon before. As you go, you'll lose stars behind you as they drop over the 'hill'. If you're north or south of the equator, there will be parts of the sky that you can never see. The 'hill' of the Earth will always be in your way.

East and west

If you travel east or west, it doesn't make any difference at all, because the Earth is turning you that way, anyway, constantly bringing you new stars over your eastern horizon and taking them away from you over the western one. But if you travel to the north or to the south, you will very obviously alter your perspective on the sky.

Latitude

Broome is 18 degrees around the curve of the Earth towards the South Pole, from the equator. From the equator to the South Pole is 90 degrees, so 18 degrees means Broome is not really that far from the equator. If you were in London you could say, 'London is 52 degrees around the curve of the Earth towards the North Pole from the equator.' Apply this to wherever you are.

Leading with your feet

Standing in Broome, you'd be tipped over at an angle 18 degrees different than when you were standing at the equator. Of course, it will look to us like we are still standing up straight because that's what we do here on Earth. Our feet will always point to the centre of the Earth no matter where we are on this round planet. People in different locations on Earth are all standing at different angles in relation to each other and also in relation to the stars.

How this translates for us in Broome is that our arms, still indicating the axis of the Earth, will now be tilted at 18 degrees. One arm is pointing 18 degrees up into the southern sky and the other one is pointing 18 degrees below our northern horizon. This is the axis of the Earth as seen from Broome.

An understanding of this axis of rotation is pretty much the key to understanding everything else that is going on in the sky.

With a working knowledge of this axis of rotation, you can go on to teach yourself all of the rest. If you play with this fundamental axis as much as you can, day after day, week after week, month after month and year after year, pretty soon you'll know. It won't be theory for you anymore. It will be a part of your reality. Bust out and do it for your whole life and then do it some more.

Celestial poles

What it means for us if we are in Broome is that, 18 degrees above the southern horizon, there is a point of rotation in the sky. We call this point the south celestial pole. The word celestial simply means 'in the sky'. So this is the South Pole in the sky, which is also a point directly above the South Pole on the Earth.

How would you like to come for a walk with me to the South Pole? Let's go!

South Pole tripping – a quick detour . . .

If we leave from Broome, we'll have 285,000 square km of sand dunes to cross in the Great Sandy Desert, then the much smaller Little Sandy Desert and the gold mines of Kalgoorlie, before we eventually reach the Western Australian south coast near Esperance and have to start swimming.

We're going all the way around the curve of the Earth to the inland of the Antarctic continent. On this epic journey to the South Pole, all of the stars will have appeared to move considerably north, until this point of rotation that started at 18 degrees above our horizon is directly overhead.

So here we are, standing at the South Pole. The Earth is turning us around on the spot, clockwise when we look down, once a day. And the stars will appear to whirl around our heads in a circle centred on directly overhead, once a day – and again, clockwise, but only because we are now looking up instead of down.

There we go, participating in and altering our own reality by simply looking the other way once again! You are a rebel! The stars near to the horizon will appear to go around the horizon, staying exactly level with it all the way around. Now you want to go to the South Pole and see it for yourselves, don't you!?

To do the same exercise at the North Pole, simply tip yourself upside down. Take your time and reverse everything in the last paragraphs and then remind yourself that the only thing that will have changed is you. The Earth and the stars will still be doing exactly what they were doing before. Pretty cool, huh!

. . . And back to celestial poles

Well, from wherever you are on Earth, you do get to see some of this effect. Looking at your point of rotation in the sky, whether you're out under the stars with me in Broome or whether you're anywhere else in the world where you can at least see stars, with the Earth turning you to the left if you're facing south, and to the right if you're facing north, the stars you'll see in those directions will be appearing to turn the other way.

We'll find the celestial poles with some more precision in the next chapter. For now, though, using the Southern Cross in the south, take note of its angle in relation to the ground. Is it standing up? Or lying on its left or right side? Is it upside down or even below the horizon where it sometimes goes in Broome? If you can find them using a star chart or an app, Canopus, Achernar, Peacock and the Pointers are the brightest stars anywhere near the south celestial pole.

If you don't know their names, it doesn't matter: simply pick a star on the other side of south from the Southern Cross. You can use the compass on your smartphone to find south if you need to. Notice how high the Southern Cross is and whatever star you have selected is and check on them 20 or 30 minutes later and you will easily notice them turning around the south celestial pole in relation to each other.

You can do the same in the northern hemisphere with the north celestial pole, using The Great Bear, and Cassiopeia, Vega and Capella, or anything else you can find straddling north.

So everything I've just taken you through has been to simply set you up to watch the Earth turn. Give yourself 20 minutes, an hour, or a couple of hours, and then check on your selections in the west, east, out (formerly known as 'up'), north and south, and you will see and know that you are on a planet that is turning, and you will know the way that it's turning in space, taking you on a journey to who knows where.

Maybe if we all start looking, this journey will reveal more of itself to us. If we all do the part we came here to do, maybe our collective *ahaaaas* will open wide the doorways of understanding. Strange as it is to say, knowing the directions that we are travelling in has literally given me direction in my own life.

Chapter 19

Great Southern Skies

We are very privileged that our spaceship Earth is fitted with a clear window to look out of. The synchronicity of our species living on a planet with an atmosphere so transparent that our human eyes can see all the way through it to inter-planetary, interstellar and intergalactic space is something to marvel at.

If we lived on the planet next door, we'd get no such view. Because Venus is permanently blanketed by thick cloud, we wouldn't know the stars and planets existed. We'd never even see the Sun! A Venetian vista would be limited to as far as the eye can see, about 5 km to the horizon – the same as here on our planet.

On a cloudless, moonless night here on Earth, however, it's possible to see the Andromeda Galaxy with the naked eye – nearly three million light years away. This remarkable range and clarity is possible in many parts of the world, but particu-larly so in the vast outback areas of Australia. The remote and unpolluted skies of my home patch in the wild Kimberley region provide bona-fide world-class stargazing conditions.

Another cool thing about stargazing in Broome – and in the tropics in general – is we have access to all of the southern hemisphere sky, and most of the northern hemisphere sky too. We can see everything that can be viewed from Perth, Melbourne, Sydney and the Gold Coast, and most of what you can see from London, Amsterdam and New York as well.

It's the southern hemisphere skies, however, that have all the treasures when it comes to stargazing. You can see the most wonderful things in the south that simply don't appear when you move north of the equator, including the globular star clusters Omega Centauri and 47 Tucanae. These astronomical marvels are by far the most spectacular clusters we can see, yet they're so deep in the southern sky they might as well not exist to stargazing Europeans. In fact, 47 Tucanae wasn't even catalogued by European astronomers until the 1750s.

The south is home to two visible galaxies as well: the Magellanic Clouds, which are a pair of dwarf galaxies that can be seen quite easily with the naked eye. There is simply no equivalent of these in the northern hemisphere. We also have access to the brightest part of the Milky Way.

The Scorpio/Sagittarius region shows us the central bulge of the Milky Way galaxy in all its obvious glory. It rarely gets high enough above the horizon in the northern hemisphere to see it at all. The Eta Carinae region to the west of the Southern Cross is an incredibly rich region of nebulosity and is peppered with enough star clusters to keep you busy with a pair of binoculars all night long. You won't see it at all from the northern hemisphere.

The northern stretch of the Milky Way through Aquila (the eagle), Cygnus (the swan), Cassiopeia and Auriga (the charioteer) is faint, even when we see it from our pristine dark Kimberley skies. Watch that same part of the Milky Way from light-polluted Europe or the USA and you might not even know that it exists at all.

In much of Australia you can roll a swag out and stare straight through the window of the sky into the very heart of our home galaxy. You can clearly discern the central bulge of the Milky Way right overhead and see where it tapers away in both directions to form the classic 'fried egg' shape.

This is just how things are. As the Earth spins on its axis, the southern half of the planet is forever orientated towards this intriguing, action-packed part of space. New Zealand, Papua New Guinea, most of South America and countries in the southern part of Indonesia and the African continent all have access to the same skies that we do in Australia. Like us, many of these nations celebrate this fortune by adorning their national flag with the Southern Cross.

The Southern Cross

The previous chapter gave you a handle on the points of the compass, your place on Earth and the direction of 'out', so now is a good time to look to the south and see what's going on. You've picked out marker stars in the west, in the east and overhead and we played a little with the up and down and round and round action that is going on around the celestial poles. Getting a little more specific, and, depending on the time of year, we're going to look for the constellation of the Crux, aka the Southern Cross.

It's one of the most prominent constellations in the southern sky, but to help us make sure we've got the right 'cross', there are two pointer stars – Alpha Centauri and Beta Centauri – which come in very handy. They're called 'pointers' because together they indicate a line that points to the Southern Cross.

If you're stargazing in the middle of the year during the southern winter months, in the civilised hours of the early evening, the Southern Cross will be standing up straight, high in the sky. The pointers will be to the left.

Three months later, as spring rolls around, the Southern Cross will be lying flat on its right-hand side with the pointers above it. Christmas and New Year will have it upside down and low in the southern sky. Depending on how far north you are, it might even be below the horizon and out of sight, like it is in Broome at that time of year.

Another three months brings autumn and it is climbing into the south-eastern sky and lying flat on its left-hand side, with Alpha and Beta Centauri pointing at it from below. And then we're back to the standing up straight and high in the sky that lets you know you've been watching the Southern Cross for exactly a year.

If you're looking for the Southern Cross from the northern hemisphere, I recommend a holiday in Australia because you won't see it at all, unless you are well south of the Tropic of Cancer. Even then you'll only get glimpses of it low on your southern horizon in the middle of the year.

Other prominent features in this deep southern region are the bright stars Achernar, Canopus and Peacock. They are all a similar distance from the south celestial pole as the Southern Cross and bright enough to be easily seen. They are great indicators for watching ourselves turn around this very special point in the sky.

Navigation

The south celestial pole itself is in a very blank section of the sky. There are no bright stars close enough to show us where it is. Finding it is indeed a very useful thing to be able to do. If you can locate this point in the sky, you can navigate. I've heard lots of ways of finding it and a lot of the ways I've heard are likely to get you lost.

I think the problem is that often someone is trying to remember how to do it. I don't know about you, but when I was in school, if someone told me I had to remember something, I didn't do very well at all. But if someone showed me how something worked, I got it every time!

So, I want to show you how to find the south celestial pole. The first part is to let you know that this point, the south celestial pole, is fixed amongst the stars. If this point is fixed amongst the stars, then all you have to do is form a relationship between this point and any stars you choose.

It doesn't matter which ones you choose, but it just so happens that if you draw a line through the long axis of the Southern Cross, through the long stick if it is a kite, extend that line away from the bottom of the kite shape and it will pass through the south celestial pole.

That is pretty handy. Then draw another line that splits the two Pointer stars through the middle in a perpendicular bisector. Where those two lines meet is very close to the south celestial pole! You can watch everything appear to move around this very special point of rotation in the sky and make it a part of your life.

Now that you can find it, see if you can find any other obvious relationships between stars in the southern sky and the south celestial pole. I bet you can, but here are a couple of examples:

If they are both above the horizon, draw a line from Beta Centauri to Achernar. The south celestial pole is close enough to the middle of the two. Check this with your Southern Cross and pointer method and see if it checks out.

You can make an equilateral triangle with Achernar and Canopus, with the invisible south celestial pole as the third point of the triangle. Peacock and Achernar pretty much make a similar equilateral triangle. These alternatives are quite handy when the Southern Cross is below the horizon and so not available to help you in your navigation.

You can make up as many other relationships with this point as you like, with any other stars that you choose, such as: 'It's this far from the star Alpha Trianguli Australis and it makes a triangle with this other star called Miaplacidus.'

It's a simple matter of realising that you're looking for that particular point in the sky. It's the same thing as finding a particular house in a town, or your favourite camping spot on a river. After a while you'll know how to find it and you will forever after. Of course, watching everything appear to rotate around this point in the sky will reinforce your knowing and cement it solidly into your reality.

Latitude in 30 seconds!

If you know how to find the south celestial pole, you can quite easily find your latitude, which is simply how far you are from the equator in degrees. One degree of latitude equals 60 nautical miles, so Broome is 18 x 60 or 1080 nm or 2000 km from the equator. Finding your latitude is easy; I can teach you how to do it from wherever you are in 30 seconds. All you need is a half-moon school protractor, a piece of string, a rock and some sticky tape. Are you ready?

Attach the string onto the zero point of the protractor with the sticky tape. Tie the rock to the other end of the string so it pulls the string straight down. Aim the flat edge of the protractor at your new-found south celestial pole, with the curved edge of the protractor pointing down. Look at where the string hangs and it will indicate on the protractor the degrees of your latitude.

There you have it! Latitude in 30 seconds.

Finding south

Now we're going to take it another step. Having found the south celestial pole using any and all of the above methods – including the ones you've made up yourself – drop a line down from this point in the sky to the ground. That is south! Now you are a navigator!

Finding the north celestial pole

Having found south and the south celestial pole, the best way to find north is to turn around! I know that sounds a little silly, but the truth is that, from the southern hemisphere, there's nothing reliable in the northern sky for finding north.

If you're in the northern hemisphere, it's a different story. It's easy. At the other end of our axis of rotation, now pointing up into the northern sky at the north celestial pole, there is a star. It's called Polaris, or the Pole Star, or simply the North Star. Finding it is as simple as drawing a line through the two stars at the end of the Big Dipper, Merak and Dubhe. They point right at it. Use your compass again if you need some help finding north.

If we're watching from the southern hemisphere, let's say from Broome, Polaris will always be below our horizon. We're rotating around that point that is 18 degrees below our northern horizon, to the right. The stars will appear to be turning the other way and so any stars that are above our northern horizon will do big sweeping arcs across the northern sky, moving left. Pick on any star in the north, drop a line to the horizon and note which tree or rock or building it lines up with. You're going to see the movement to the left fairly quickly.

Depending on the time of the year and the time of the night, you might find the bright star Arcturus out there. The Big Dipper makes an appearance early in the year if you live in the tropics. The Gemini twins, plus Vega, Altair and Deneb, are some of the other stars you find in this part of the sky.

Now you know a little more about the rotational axis, you'll be better placed to watch the Earth turn, to find north and south and to calculate your latitude on the planet in under a minute.

A brief look at longitude

Finding longitude – or your position east and west on Earth – using celestial navigation is far more difficult. It involves measuring a known star's elevation, knowing the time of the measurement, the day of the year, and calculating where that star should be at that time on that day if you're at the calculated position on the Earth's surface, at sea level.

The room for error is huge, as each calculation is critical to the accuracy of the next. You'll need a comprehensive set of tables of star positions, with dates and times and a precision clock too.

A more reliable method of finding longitude before the invention of the clock was to keep sailing east or west until you crashed into somebody's country. Western Australia regularly featured in the practice, with many shipwrecks scattered along our coast. Today, a fleet of GPS satellites orbiting the Earth twice a day provides us with known points in the sky. They triangulate with each other and with our GPS receivers (you've got one in your smart phone), to give us an accurate position on the Earth. This makes it easy and relieves a lot of stress when it comes to sailing around the world, or simply going to the shops.

The celestial equator

Dividing the two hemispheres is the celestial equator, a line in the sky directly over the equator on the Earth.

The stars and constellations that are on the celestial equator can be seen from anywhere on Earth. Orion – one of the most conspicuous and recognisable constellations of all – is a perfect example of a fixed marker on the celestial equator that works for everyone in the world. However, since billions of humans are upside down from each other depending upon which hemisphere they're in, Orion looks very different from different perspectives.

In the northern hemisphere it's known as 'the Hunter', after Orion the Hunter from Greek mythology. Its brightest stars are the blueish white Beta Orionis (Rigel) and the red Alpha Orionis (Betelgeuse). The Hunter appears standing upright and brandishing a sword. Three more bright stars – Alnitak, Alnilam and Mintaka – form a belt across his waist.

Where I come from south of the equator, Orion appears upside down and looks nothing like a man holding a sword. In fact, it looks more like a cooking pot, with a long handle and three bright stars – yes, Alnitak, Alnilam and Mintaka – as its base. This is why just about everyone in Australia knows Orion as the saucepan.

If you can't find Orion, you'll quite likely see the constellation of Scorpio instead, because Orion and Scorpio are opposite each other in the sky. As one disappears over the western horizon, the other is always appearing over the eastern one.

Equatorial constellations appear to trace straight lines across the sky moving from east to west, while stars either side of the celestial equator seem to follow curved paths across the sky. The closer we get to the poles, the tighter these circles appear to get. Now is as good a time as any to remind you that none of these stars are moving. That's us doing our Earth-turning thing again.

Circumpolar stars

From the equator you can see all of the southern hemisphere skies and all of the northern hemisphere skies. If you're above or below the equator, you'll permanently lose sight of a part of the opposite hemisphere.

From either of the poles you'll only see the stars of that hemisphere, all the way to the celestial equator. Every star you see will be permanently above your horizon and what we call 'circumpolar'.

Travel a little away from the pole and you'll begin to see over the equator and into the other hemisphere. You'll also have some of the sky in your own hemisphere spend time below the horizon as the Earth turns.

From Broome, at 18 degrees south, we don't see any of the stars that are within 18 degrees of the north celestial pole. This is a relatively small section of sky. However, the stars within 18 degrees of the south celestial pole are 'circumpolar', a fancy way of saying that they go around and around in the sky without going below the horizon. The further south you go, the bigger the circle of sky that is circumpolar and the bigger the circle of the northern sky you will never see. Of course, the same thing is reversed when you go to the northern hemisphere.

Ursa Major – the Big Dipper – is near enough to 30 degrees from the north celestial pole, so it is circumpolar for most northern observers. People living in southern Australia – i.e. most Australians – don't get to see it at all, because it's permanently below their northern horizon. In Broome it crosses our northern sky beautifully in the early part of the year.

Conversely, the Southern Cross is also around 30 degrees

from the south celestial pole, making it circumpolar from southern Australia, New Zealand, the southern tips of South Africa and South America. We get it high in our Broome sky all through the best dry stargazing season, so we don't really mind that it isn't circumpolar and goes below our southern horizon during our cloudy wet season.

To see the Cross at all, you need to be less than 30 degrees north of the equator, which leaves out the UK, Europe and most of the US too.

So . . . come on down!

Chapter 20

Astronomy vs. Astrology

To recap a little, the ecliptic – the plane of our solar system – is the only part of the sky where you will ever see the planets, or the moon, or the Sun. The zodiac constellations marking it out – Aries, Taurus, Gemini, Cancer, Leo, Virgo, Libra, Scorpio, Sagittarius, Capricorn, Aquarius and Pisces – are all very real territories in our skies.

Whether you're into horoscopes and 'star signs' or whether you reckon all that stuff is hocus-pocus, in occupying the ecliptic they clearly show us the plane of the solar system and help us to know we're looking in the right place for the planets.

As the Earth goes around the Sun, the Sun appears to shift its position against the backdrop of particular constellations at particular times of the year. When it is Gemini time, the Sun is in front of the constellation of Gemini from our perspective. In other words, Gemini – or at least an aspect of Gemini – is hidden behind the Sun. This means Gemini is in our daytime sky. The Earth moves a bit further around the Sun and the Sun appears in front of Cancer, Leo, Virgo and so on.

Two circles

The ecliptic is not the celestial equator. You can't see either of these two circles drawn on the sky, but you can learn where they are. Because the Earth is tipped over at 23½ degrees in relation to its journey around the Sun, the ecliptic – or the plane of the solar system – from our perspective here on Earth, is tipped over in relation to the Earth's celestial poles and to the celestial equator.

How this looks for us in the sky is that half of the ecliptic is below the celestial equator and half is above. Half is in the north and half is in the south. The two circles of the celestial equator and the ecliptic cross each other in the sky at two points that we call equinoxes. These are the points where the Sun appears to cross from the northern to the southern hemisphere, and vice versa.

Equinoxes

The ecliptic crosses the celestial equator going north at a point we call the 'vernal equinox' on 20 or 21 March every year. In the northern hemisphere, this is the spring equinox, while south of the equator it is the autumn equinox. When the Sun appears at this very specific point in the sky, it is the time of equinox on Earth. Both hemispheres are equally orientated at the Sun as it appears to cross directly over the equator, heading north. The hours of day and night are equal.

The vernal equinox is the starting point, or arbitrary zero point, of the ecliptic and is also called 'the first point of Aries' by both astronomers and astrologers. (It actually currently sits somewhere on the boundary between Pisces and Aquarius, thanks to something called the 'precession of the equinoxes', which I'll come to a bit later.)

It's the same with the opposing equinox, when the ecliptic crosses the celestial equator going south on 22 or 23 September. It is the autumn equinox in the northern hemisphere and spring equinox in the south. This is called 'the first point of Libra', even though it currently sits somewhere on the boundary between Virgo and Leo in the sky.

Solstices

The celestial equator and the ecliptic cross each other at the equinoxes, and the times that they're furthest apart are called the solstices. They are very specific points in space – as well as being very specific points during Earth's journey around the Sun. It's when the Earth orientates one or the other of its poles as directly towards the Sun as they can be.

For six months of its solar orbit, Earth's northern hemisphere is pointed towards the Sun. The Sun is pouring down from overhead in the northern hemisphere. In the six months following that, the southern hemisphere is pointed at the Sun and the Sun is overhead somewhere between the equator and the Tropic of Capricorn.

Each solstice is either the longest day of the year or the shortest, depending on which hemisphere you are in. When the North Pole is most fully facing the Sun, the Sun sits directly over the 23½ degrees line of latitude north of the equator. This is the Tropic of Cancer. The Sun will appear to be in Cancer in the sky. Cancer is the northern-most constellation on the ecliptic.

This solstice is called the 'first point of Cancer'. Astrologers and astronomers agree on this one too, even though this point is somewhere between Gemini and Taurus. The Sun will appear to be in Cancer if we measure Cancer as being related to – and lined up with – the solstices and the equinoxes. (Again, this relates to the precession of the equinoxes, which I'll get to soon.)

When the South Pole is most fully orientated towards the Sun, the Sun sits directly over the 23½ degrees line of latitude south of the equator. This is the Tropic of Capricorn. The Sun will appear to be in Capricorn in the sky, as we measure Capricorn from this solstice. Capricorn is the southern-most sign of the ecliptic. This solstice is also called the 'first point of Capricorn'.

Wobbly ecliptic

Having half of the ecliptic above the celestial equator and half of it below means that as the Earth turns for you during a night of stargazing, the ecliptic's position north or south of the celestial equator varies too. This means it varies in height above your head during the course of the night and during the course of the year.

The celestial equator stays constant in its position in the sky as long as we stay constant in our position on Earth. It splits the sky between the two celestial poles. The ecliptic appears to sweep backwards and forwards across the sky 23½ degrees north and 23½ degrees south of the celestial equator, if you stay up watching it all night. It will do the same thing during the course of the year if you decide to watch it at the same time each night.

Learning the zodiac constellations will help you track this fascinating apparent motion.

I'm going to remind you again that all of this movement is that of the Earth spinning. Everything I address here – wobbly ecliptic and all – relates to our changing view from our ever-turning planet. Keep this in mind and everything will become so much easier. The fact the Earth is tipped over at 23½ degrees while it spins just makes our lives and our observations of the ecliptic more interesting.

Broome, romancing the ecliptic

Broome has a wonderful relationship with the ecliptic. It runs very high across our sky, particularly during our peak viewing time in the dry season from April to November. Crystal-clear, star-filled nights are quite normal for months at a time. Balmy, comfortable night-time temperatures allow for hours of viewing pleasure too.

Having the ecliptic high in the sky means the planets have to cross through the highest part of the sky, where we get to see them through the least amount of the Earth's atmosphere. This means clear, sharp views. It also means the planets can be very high in the sky for many hours of the night, allowing a lot of time to study our planetary neighbourhood.

In the southern states of Australia, the best viewing weather is in summer when the ecliptic and planets are low down across the northern horizon at night. At best you'll get an hour or so of viewing the planets in a high enough part of the sky for them to be clear. During winter, most people are indoors with the fire going, although some of those crisp, clear, southern winter nights do provide spectacular stargazing. If you can brave the cold, the ecliptic will be higher in your northern sky, making viewing the planets a much more viable proposition.

It's the same situation for stargazers higher up in the northern hemisphere. The ecliptic is low on the southern horizon during the summer evenings, with the planets grazing along the southern horizon at best. In winter, Europeans, Americans et al face the same obstacles as their southern hemisphere cousins.

Legends and ritual

The solstices and equinoxes are four specific times of the year. They are also four specific places in the very fabric of space the Earth travels through each year. Equinoxes and solstices are the stuff of druids and full moon rituals, of drunken debauchery and midnight romps.

Why is it that these times that were once glorified and celebrated are, more often than not, ignored today? One reason could be that we spend most of our nights indoors in front of the TV and rarely, if ever, tune ourselves in to the world around us. Perhaps the ancients had a little more time on their hands.

Whatever the case, it's possible for us to tune ourselves in to these turning points in the Earth's journey around the Sun with some simple observations carried out during the year. These are natural switching points – times at which the Earth has distinct changes of direction. The Sun 'clicks' over the equator at the equinox and 'pushes the buttons' of the solstice, before reversing its direction and our relationship to it. The Earth is side-on to the Sun at the equinox and points one pole or the other clearly at the Sun during the solstices.

Traditionally these have been times of reassessment, of tuning in to what we are doing in our lives and deciding what we want to keep doing and what we want to change. Perhaps the solstices and equinoxes can become practical tools in your life with a simple acknowledgement of their existence.

Notice the new shoots springing out of the ground on 22 September at the southern spring equinox or on 21 March at the northern spring equinox. The four seasons of the temperate world are triggered by the solstices and equinoxes. Maybe just noticing how you feel and what's going on in your life will help you make good decisions at these very powerful and very real points in the Earth's annual journey around the Sun.

Precession and other mysteries

Something that's a little harder to tune in to is the precession of the equinoxes – simply because it takes a very long time to complete one cycle. The precession refers to the movement of the solstices and equinoxes in relation to the stars. The vernal equinox – the first point of Aries – travels all the way around the sky and through all of the zodiac constellations, going backwards, once every 25,920 years.

Two zodiacs

This leads to a fair degree of confusion for most people – even among those who study these things. There are two zodiacs:

the zodiac constellations you see clearly in the sky as stars, and the 'zodiac signs' as determined by the solstices and equinoxes.

As the solstices and equinoxes shift in relation to the stars due to precession, so do the zodiac signs. You can't see the zodiac signs unless you have already figured out where the solstices and equinoxes are in the sky. There are no obvious stars marking any of these cosmic territories.

The solstices and equinoxes are not visible unless you look for the signs that show us that they are happening. Stonehenge, the pyramids, the Jantar Mantar at Jaipur in Rajasthan and any other number of ancient solar observatories were constructed to make it easier for us to see these very real places in the sky, and to acknowledge the very real positions in space that the Earth travels through on these very special days. The Sun lining up between two points on Earth on the same day each year tells us that we're in a particular position. We've plotted enough of these positions over the millennia to now call them solstices and equinoxes. This is how we've been able to plot the precession of the equinoxes too.

'Precession' has got a lot to answer for. It creates a situation where the visible 'constellations' of the modern and ancient astronomers don't match up with the invisible 'signs' of the ancient and modern astrologers. Well, they do match up, but only once every 25,920 years or so.

The most common explanation for precession is that the Earth is wobbling around on its rotational axis like a spinning top winding down and doing a wobble once every 25,920 years. Maybe this is how it works and perhaps it's something else entirely. Maybe a nearby dark star that we haven't been able to see yet gravitationally affects the Earth's orbit. We don't know. I quite like it that way.

What is clear, though, is that the position of the equinox – where the ecliptic crosses the celestial equator – moves slowly backwards around the zodiac constellations. Put another way, the Earth comes to a slightly different position on its journey around the Sun to achieve equinox every year.

This is the dawning . . .

This is how we are coming out of the 2160 or so years of the 'Age of Pisces' and moving into the new 'Age of Aquarius'. The first point of Aries – the vernal equinox – or the exact point in the sky that the Sun appears to be when it crosses the celestial equator moving north, is shifting from the visible constellation of Pisces and into the visible constellation of Aquarius right now.

The exact date for this change is hard to pinpoint due to the occasional rearrangement of constellation boundaries throughout history. It's also a very slow switch as the equinox moves only about 0.014 of a degree each year, or 1 degree every 72 years. So . . . watch this space!

Astronomy and astrology

Modern astronomers love to bag modern and ancient astrologers because of the difference between the positions of the zodiac signs and the zodiac constellations.

'Astrology is a load of rubbish because astrologers say the Sun is in Gemini when it's really in Taurus,' is a common statement you'll find in the astronomy textbooks that care to comment at all. It seems to me there is at least as much ignorance amongst astronomers as there is hocus-pocus amongst astrologers. In reality, both systems have merit and both systems offer something different.

Astronomy is a study of the physical realities of existence. Astrology is a study of cycles and the patterns that emerge out of putting many cycles together. So much about the cosmos is cyclical: the Earth's daily rotation and its annual jaunt around the Sun, the moon's month-long repeat trip around Earth, Mercury's 88-day cycle, Venus's 7½ months, Mars's two years, Jupiter's 12 years and so on. When you observe enough of these cycles, you'll see patterns emerging from them as the planets pass each other in the sky in a regular fashion.

Astrology and its interpretation of our emotional and spiritual states is something I'm not yet qualified to comment on. Give me another 500 years or so and we'll see how I go then. There's that 'wait and see' approach again. I'm a patient man. A simple observation of your own reactions to the full moon may well provide you with some food for thought on the subject.

Predicting the future is something astrologers are often challenged on by science. Yet astronomers also predict the future every time they post sunrise and sunset times or draw up a tide chart. We are so familiar with the astronomical predictions of these simple cycles in our daily lives, and we've seen them come true so many times, that we accept them as gospel truth.

If you study the daily newspapers as a case for astrology, I'm sure anyone could soon dispute what is found there. If you were to consult with a learned astrologer, however, I'm equally sure you would find more than a few insights.

Esoteric terms

As to the argument about which zodiac sign belongs where in the sky, astrologers divide the sky into 12 equal segments – the zodiac signs – that start at the vernal equinox, or the first point of Aries. There are 12 signs of 30 degrees making up the full 360-degree circle.

Astronomers divide the sky in exactly the same way – into 24 equal segments of 15 degrees each, making the full circle. They call these the 24 hours of 'right ascension'. They also start off their divisions from zero at the vernal equinox and they too call this the first point of Aries.

The systems are identical and shift in relation to the zodiac constellations according to the precession of the equinoxes. The first and second hours of right ascension are Aries, the third and fourth are Taurus, and so on.

Both sides of the story use the same starting point for their coordinate systems and give it the same name. The only difference is the names of their divisions. The 12 'zodiac signs' and 24 'hours of right ascension'. Which name sounds more esoteric or 'out there' to you?

Chapter 21

Moon Nodes and Eclipses

The orbit of the moon is inclined to the ecliptic by about 5 degrees. During a lunar month the moon spends half of the time north of the ecliptic and half of the time south of it.

This is slightly reminiscent of the relation between the celestial equator and the ecliptic from the last chapter, only this time it is a relation between the ecliptic and the moon's orbit. The 'moon's nodes' are the points in the moon's inclined orbit where it crosses the ecliptic, either going north or going south, a bit like the equinoxes, only this time specific to the moon.

If the moon's orbit was aligned with the ecliptic, there would be no nodes, no movement north or south of the ecliptic, and the Sun and the moon would line up every time the moon went around the Earth. We would have a lunar eclipse at every full moon and a solar eclipse at every new moon.

As it is, the only time we get an eclipse of any sort is when the new moon or the full moon occur at the same time the moon is passing through one of its nodes. A full moon brings a lunar eclipse as the moon passes through the shadow of the Earth, and a new moon will cast the shadow of the moon onto the surface of the Earth as a solar eclipse. If you're lucky enough to be standing under the shadow of the moon, you will witness one of the most amazing sights on Earth where the moon blocks the Sun out of the sky.

Solar eclipse

By an amazing coincidence, the moon is 400 times smaller than the Sun and the Sun is 400 times further away than the moon, making them appear to be the same size in the sky.

One or two solar eclipses happen somewhere on Earth every year. To see one, you'll need to travel to the path on the Earth where it is happening. Unless you're in the right place you won't see anything at all. The moon's ability to block the Sun out of the sky completely is an amazing phenomenon that you really must witness at least once in your life. If you're ever anywhere near a solar eclipse, go to it, even if it means travelling out of your way. It is one of the most powerful experiences you will ever have.

In December 2002, I loaded up my old LandCruiser with a few telescopes for a trip to Ceduna on South Australia's Great Australian Bight. We drove 5000 km to be in Ceduna for 33 seconds – that's how long the moon completely blocked out the daytime Sun for. Yes, it was well worth the effort!

We arrived from Broome with about an hour to spare before the moon started its assault across the face of the Sun. We set up an 8-inch Celestron Schmidt-Cassegrain telescope with a full aperture solar filter for a fully filtered direct viewing of the Sun. The Sun appears orange in these filters and the sunspots show up beautifully. Fairly soon we could see the moon starting to block out the Sun, taking its time to creep across and take a bite out of the burning disc.

We'd picked a site on the side of a hill next to a freshly harvested wheat field, 10 km inland from Ceduna. The shadows on the ground grew into crescents, the light dimmed and gained an eerie quality, and then it was dark.

The solar filter came off the Celestron and the view of the solar corona through the 8-inch scope, with purple filaments of flame leaping off the Sun's surface, is something I'll remember forever. Looking up at the blocked-out Sun in a darkened sky hit me so hard it felt like a physical blow. I can only describe it as an intense spiritual experience that had me crying out loud with the power of it. All too soon, though, our 33 seconds of totality was over and the moon continued on its journey around the Earth, slipping off the face of the Sun as it went.

After that rushed trip from Broome to get to Ceduna on time, we meandered back across the cliffs of the Great Australian Bight, exploring the many limestone caves along the way. We then headed for Rottnest Island for a season of summer Astro Tours. I was looking forward to furthering my surfing career at the Strickland Bay surf break at the same time.

You need to travel for a solar eclipse because the moon casts only a very narrow shadow on the Earth. Unless you're on that track of the moon's shadow – which is 280 km wide at the most – you will miss out.

Stand near the edge of the shadow line and you'll get a partial solar eclipse with the Sun only partially blocked out by the moon. Totality is about a million times better than a partial eclipse, so do whatever you have to do – walk, crawl, run, drive, fly, swim – to get yourself in position directly under the shadow of the moon.

I met people in Ceduna who, having chanced upon one eclipse in their life, have since dedicated their time to travelling all over the world following them. I can well understand why. I've seen three so far and missed another one due to cloud. At the time of writing, I'm looking forward to a 1400 km drive to Exmouth from Broome for the 20 April 2023 Hybrid Solar Eclipse, to stand in one minute and one second of darkness as the moon blocks out the Sun.

Five years later, on 22 July 2028, I'll be doing the trek deep into the heart of the wild Kimberley Plateau for five minutes and 10 seconds of darkness. The 230 km wide shadow of the moon crosses Australia all the way from the magnificent Kimberley coast to a direct hit on the heart of Sydney.

Lunar eclipse

A lunar eclipse is much easier to see. All you have to do is be on the night-time side of the Earth when it is happening. You won't even have to leave home. The first time I saw one, while reclining on my motorbike at the top of a mountain outside of Alice Springs in the early '80s, I had no idea it was going to happen. Nowadays you can easily look up the next scheduled eclipse on the internet.

A lunar eclipse happens when the moon crosses one of its nodes at the same time as the full moon. The full moon passes through the shadow of the Earth for anywhere up to a couple of hours. If the moon passes through the deepest part of the

Earth's shadow, the umbra, it is likely to dim considerably. It usually won't go completely black, as sunlight bends around through the Earth's atmosphere to light up the moon with deep reds and browns. At least two – and as many as five – lunar eclipses occur every year.

The moon travels halfway around the Earth from one node to the other in about 14 days, so solar and lunar eclipses almost always follow each other two weeks apart, and then six months later again when everything lines up with the other node.

The moon's node shifts backwards by about 20 degrees against the ecliptic every year. In effect, this means that the inclination of the moon's orbit to the ecliptic has an 18½ year backwards wobble around the ecliptic. Because of this, eclipses happen about 11 days earlier every year and shift constellations backwards around the zodiac as they go.

What an amazing universe we live in!

Chapter 22

The Moon and the Tides

Although I nowadays make a living out of gazing into space, the oceans of planet Earth are a constant reminder of where that fascination began. With the ocean level rising and falling 10 m in a six-hour period for Broome and 12 m just up the road in Derby, the Kimberley coast has the biggest tides in the southern hemisphere and the second-biggest tides in the world. Imagine the whole surface of the ocean rising and falling 10 or 12 m! Where does all of that water go? I still wonder exactly that every time I go down to the 22 km of white fresh tidally washed sand that is Broome's famous Cable Beach.

The rhythmic ebb and flow of our incredibly powerful and obvious tides played a huge role in getting me interested in the sky and I still think of the instructive maps one of the Kimberley's great gentlemen, Jeff Johnson, scratched into the mud on the side of a street in Broome for me in early 1982.

I had just arrived in town and picked up work as a pearl diver when Jeff took it upon himself to talk me through the lunar cycle. He explained enough about how the moon and the tides work together for me to go off and figure out some more for myself. I was working to the rhythm of the tides every day, so I had the perfect opportunity to test out my new knowledge.

What is a tide?

The high tide is simply a bulge of water that's pulled up underneath the moon by the sheer force of its gravity. In fact, that gravity attracts everything – even the solid Earth, which flexes in response. Of course being fluid, the oceans move a lot more than the relatively solid outer crust of the planet.

When the moon is overhead, it's high tide. When the moon is on the horizon, it's low tide. It's as simple as that, although the time of high tide varies with a lag of an hour or so, depending on your location on Earth. River mouths, bays, headlands, currents, marine trenches and other shapes and contours of the ocean floor will all affect the size and the time of the tide wherever you happen to be.

Earth spins on its axis much faster than the moon goes around the Earth, so when it comes to the daily tides, the moon stays relatively still while the planet turns your part of the world through the bulge of water that is pulled up under the moon. This high-tide bulge behaves like a wave that moves all the way around the planet once a day, as it spins from west to east.

Doubling up

The tug of the moon only explains one of the high tides that we get every day. There are two high tides and two low tides per day. This varies too with your particular location on Earth and the many varied factors that apply to the tides where you live. Four tides a day, however, is a classic tide scenario and that's what we get in Broome.

We talk about the moon going around the Earth when, in fact, the Earth and moon wobble around each other, orbiting a common centre of gravity. If you managed to weigh the Earth and moon together and had to balance them, you would find the balance point a couple of thousand kilometres below the Earth's surface. The centre of the Earth is more than 6000 km below your feet right now! The Earth and the moon are wobbling around this common centre of gravity.

Two forces, two high tides

The Earth and the moon are pulling towards each other with the gravity each of them has. They are also trying to get away from each other with the centrifugal force of their rotation around each other. As the moon pulls on Earth to keep the relationship going, it also pulls the bulge of water that is one of the high tides. At the same time, the centrifugal force of the Earth and moon going around each other throws a bulge of water out the opposite side of the Earth. There are two high tides happening at the same time, on opposite sides of the planet.

The moon's gravity has less effect on the opposite side of the Earth too, simply because the moon is further away. This means the moon's gravity offers less opposition to the centrifugal force that throws the bulge of high-tide water out on the opposite side of the globe.

The water pulled up in the tide that's under the moon's gravity also has centrifugal force, which throws it even higher because the centre of spin is underneath it, below the surface of the Earth. This explanation doesn't include all of the factors involved, but somewhere between these two effects of gravity and centrifugal force, we get a high tide on opposite sides of the Earth at the same time.

Two low tides

Between these bulges of high-tide water – which are just over 12 hours apart – are the low tides. The moon will be on either horizon. As the moon appears to shift east across the sky by nearly an hour every day, so, too, the tides are nearly an hour later every day.

Some places in the world have only two tides a day, one in and one out. In some places, the incoming and outgoing tides meet each other, cancel each other out, and the result is very little tide at all. Being a wave in nature, harmonics and resonances apply to the tides, amplifying them or cutting them out altogether. On a perfectly spherical planet with no landmasses in the way and a uniform depth of ocean, the tides would behave perfectly too.

The moon phases and the tides

In the rollicking dance between the Earth, the Sun and the moon, there are times when all three are lined up. There are also times when they are not lined up at all. As we have already seen, the times of alignment are called 'new moon' and 'full moon'.

At new moon, the moon is between the Earth and the Sun. At full moon, it is opposite the Sun from our perspective on Earth.

At both of these times of alignment the Sun and the moon's gravity combine to give us bigger than usual tides. The tides rise higher and fall lower. These are known as spring tides. Cable Beach is washed clean twice a day all the way from the sand dunes to the low tide mark on what is now a 400 or 500 m wide stretch of sand. You can be the first one in the morning to put your footprints on it.

Conversely, at first quarter moon and third quarter moon, the Sun and the moon are square to each other as seen from Earth. There is no alignment and no combined gravitational pull, so the moon – still being the main driver of the tides – doesn't have the gravity assistance from the Sun. The tides are smaller. They don't rise as high or fall as low. These are known as neap tides. With a tidal movement of only 4 or 5 m in and out, Cable Beach becomes a mess of footprints for days at a time, as the water doesn't come in far enough to clean it up.

The mechanics

The moon drives the tides because even though it is small compared to the Sun, it is much closer to us. Gravity decreases with distance by the 'inverse square rule', which means that if something is 10 times further away, the gravity is 100 times weaker. Gravity is directly related to mass, in that if something is 10 times more massive, it will have 10 times the gravitational force.

The Sun is a good 27 million times more massive than the moon, but it is 400 times further away too. So 27 million divided by 400 squared still works out with the Sun having 169 times more gravitational effect on the Earth than the moon does. Confusing, right?

The next factor is the Earth and moon hurtling around the Sun fast enough that the centrifugal force holding them in a stable orbit around the Sun cancels out enough of the gravitational attraction from the Sun to allow the moon to dominate in the tide arena.

Figuring out the centrifugal force of the Earth and the moon going around the Sun means applying the inverse cubed rule. So 27-odd million divided by 400 cubed means the Sun has got 0.423 of the gravitational force of the moon as effected on Earth. In other words, the moon has 2.365 times more influence on our tides than the Sun.

These numbers seem about right looking at the tides as I see them. The numbers were never really that important to me. Being able to look at the moon and know what the tide is doing has been of far more use in my life.

Knowing how far out on the nearly 2 km of mud flats of Broome's Town Beach to anchor a 16-foot work boat so that it's floating the next day is a good use of this knowledge. Parking it too far out means swimming out to it on a cold July morning. Parking it too far in means that it will be dried out on the mud and have to be dragged across the flats to get it floating to go to work. I got to the stage where I could look at the moon and know where the tide was going to be that day; how big it would be and at what time. I got good at it simply because I was using this information every day.

Stairway to the moon

Broome's ridiculously clear skies let us see the moon go through its regular cycle, and our 10 m tides are big enough to play a significant part in the whole community. Broome people celebrate this relationship between moon and the tides every month by gathering to witness a beautiful phenomenon called the 'Stairway to the moon', which happens two or three days immediately following the full moon. Around this time:

- The tides are always big. These are the 'spring tides'.
- The huge high tide that covers the mud flats of Roebuck Bay is always at lunchtime.
- Low tide is always at sunset with the tide way, way out.
- The exposed low-tide-rippled mud flats are glistening wet from the high tide at lunchtime.
- Moonrise is always at sunset on the full moon. By definition, a full moon is an alignment of the Sun, the Earth and the moon. You are standing in the middle. At full moon time, you can watch the moon rise as the Sun sets.
- A day after full moon, the moon rises just after the Sun sets. The next night it is later again by 50 minutes to an hour. On both nights the sky is dark enough to see the moon and its reflection in the wet mud flats.
- The rising moon, appearing across the east-facing Roebuck Bay, reflects off the wet rippled mud flats of the bay.
- There is a direct reflection of the moon in each ripple in the mud flats.

The result is a lighted stairway coming towards you across the mud flat, leading all the way across the east-facing shores of Roebuck Bay to the moon. Thousands of people line the bay each full moon to witness and celebrate this amazing natural event. Maybe we'll see you up here next time.

Chapter 23

Tools of the Trade

Just as the Sun's light prevents us from seeing the stars during the day, poorly designed outdoor lighting does much the same thing at night. Upward-facing lights scatter light through the atmosphere and thus dilute the darkness we need for optimal stargazing. In suburban settings this light pollution is often accompanied by uninvited 'light trespass', where streetlights and outdoor lights from neighbouring properties shine directly onto your property.

It's even worse in the US, the UK and Europe, where the light from cities spills from one to the next. Finding clear, dark skies in the northern hemisphere is a major challenge. Many a European visitor to Astro Tours has told me they've grown up without ever seeing a star. In Australia – a massive continent with a small population – we take for granted the stars we've all seen since childhood, even in the big cities.

A clear sky is the first essential for stargazing. Find a suitable place away from towns and cities and give yourself 20 minutes under the star-filled sky. Your eyes will reward you by letting you see it is not as dark as you thought. Even in the absence of the moon, starlight easily illuminates the landscape when the sky is clear and free from light pollution.

It takes 20 minutes in darkness for healthy pupils to open to a maximum of about 6 mm. They will close up in less than one second if you expose them to white light and it will then take you another 20 minutes to get your night vision back. That could pose a problem if you're trying to operate a telescope or consult a star chart in the dark and you reach for a torch. This is why astronomers use red or orange lights that preserve human night vision. Remember, the wider open your eyes are, the more stars you will see.

Binoculars

Instead of letting 6 mm of light into your eyes, binoculars will typically cram 50 mm of light in. Try them out by looking out on a dark, night-time landscape; you'll be amazed how much more clearly you can see. Point them at the sky, however, and you'll see hundreds more stars in places where you thought there were none. If binoculars have a 50 mm front lens, or aperture, this effectively means your pupils are 50 mm wide. Telescopes and specialist binoculars have even bigger apertures that allow you to see fainter and more distant things.

A far less important feature of a telescope or binoculars is magnification. People often ask which of my telescopes is the most powerful. Usually they're asking about the magnification and want to know if we are effectively 100 or 500 times closer to the object we're looking at.

Cheap and nasty telescopes exploit this and we used to see them advertised as having magnification of 500 times or more. What they didn't mention was how a cheap telescope won't let you see anything worth looking at. When you magnify a blurry image, it gets even more blurry, even in professional quality telescopes. The real measure of a telescope is its aperture or size. The bigger the telescope, the clearer and brighter the images will be.

Binoculars typically magnify the image seven, eight or 10 times and this is plenty good enough to show you far more than your own eyes, which will show you things at one-times power. Believe me, one-times, or 1x, is still my favourite sky-viewing option. Some specialist binocular sizes jump to 20x or more. Most telescope viewing is done with less than 100x magnification, allowing wide-angle objects like star clusters, globular clusters, nebulae and galaxies to fit into the narrow field of view of most telescopes.

Higher powers are used mainly for the planets and for splitting double stars. Particularly clear nights with a steady atmosphere may let you explore the realms of 400x occasionally, although image quality often suffers. If you start with a clear, sharp image in the telescope at low power and then magnify it, you will also magnify any flaws in the image.

Changing the magnification of a telescope is as easy as changing the eyepiece. They slide out of the focus mechanism to be replaced with eyepieces of different focal lengths to achieve different magnification. The telescope aperture, or opening, collects the light and focuses it to a point where the image is magnified by the eyepiece, allowing you to see it up close. Eyepieces vary greatly in focal length, quality, field of view and price.

Binoculars are still my favourite stargazing tool. They'll show you hundreds more stars than you can see with your own eyes. If you've got a pair already, get them out tonight and point them at the sky. Trust me, it'll blow your mind. Even if you do go on to buy a telescope, you'll still need a pair of binoculars to find the things in the sky you want to check out with the telescope.

The first number on a pair of binoculars is the magnification, while the second number refers to the aperture. So 10x50s magnify 10 times with a 50 mm aperture. Binoculars are usually fixed in their magnification, although zoom models are available. The maximum magnification you want for hand-held astronomy is 10x, although 7x are the easiest to hold. An increase in magnification increases the hand shakiness too. Anything above 10x would require you to mount them on a tripod.

A 50 mm aperture is desirable for its good light-gathering power. However, 8x42 has become my favourite size as they are a little smaller and lighter, making them easier to pack on my motorcycle. With good technology optics, they perform almost as well as a pair of 10x50s.

Small aperture bird-watching binoculars aren't really suitable for the sky, as their 20 or 25 mm aperture results in images that are too dim. A good pair of wide-angle 10x50s is the best thing in the world for scanning around the sky and discovering hundreds of wonderful sights. Learning the sky with a simple star chart and a pair of binoculars is a great formula for years of joy, before you even need to think about a telescope. This is what I learned from my *Penguin Dictionary of Astronomy* all those years ago. It was good advice!

Telescopes

Lots of people buy a telescope, get it out twice, find a couple of stars to look at, realise that it's not that easy to find pretty things in the sky, and then put the new toy in the cupboard to gather dust. After a couple of years exploring with binoculars, you'll have found lots of objects you want to see close up.

You'll know a bit more about telescopes too and are more likely to have bought the one that suits your purposes. For most buyers, telescopes come in two main types – refractors and reflectors.

The refractor

These classic look-in-one-end-and-out-the-other telescopes use glass lenses that collect and focus light to a point. Glass lenses bend light and, just as a prism splits light into its rainbow of colours, a glass lens focuses different wavelengths of light, or different colours, at different distances. This is an inherent problem with a refractor as it creates false colour images in your telescope.

Cheap refractors usually have a lot of this colour aberration, which leads to poor quality images. Quality refractors minimise colour aberration with the use of multiple lenses and high-quality fluorite and exotic glasses. They are expensive, although well worth the investment, with high contrast making them excellent for viewing planets.

The reflector

Instead of passing light through a glass lens to bring it to focus, reflectors use a curved mirror to collect and reflect light back to a focus in front of the mirror. A secondary mirror catches the collected light and throws it out to the side or back down a hole in the main mirror, so you can actually get your eye to it.

Mirrors are cheaper to make and have no problems with colour aberration. Big aperture telescopes become quite afford-able, especially in the classic Newtonian design, which uses a hand-ground parabolic curved mirror to achieve focus.

Schmidt-Cassegrain telescopes use a spherical mirror, which can be machine ground, allowing mass production. Spherical mirrors have their own optical problem with spherical aberra-tion, which is corrected in a Schmidt-Cassegrain telescope by putting a correcting lens in the light path in front of the mirror.

These compact telescopes have a short fat tube and are easily packed into a foam-lined trunk for transport. One of my 8-inch Schmidt-Cassegrain trunks is decidedly beaten up from too many exploration treks along the Kimberley's mighty Fitzroy River tied on the back of a camel. Brushing against trees to give themselves a scratch is a favourite thing for a camel and is a really good test for a foam-lined telescope trunk.

Newtonian telescopes with the same aperture have a much longer tube, making transporting them an issue, particularly when they are big! However, if you're not moving them too often, the best value for money and the best value in terms of ease of use is a Newtonian telescope mounted in a Dobsonian mount. They are by far my favourite telescopes for ease of use and excellent bang for your buck. We use them in apertures from 10 inches to 20 inches at my Broome starshows. These are seriously big telescopes 20 inches or 500 mm in diameter, and standing over 2 m tall.

There are lots of other telescope designs with many

variations. Large aperture telescopes work best under dark clear skies. They are less use in the city, due to their tendency to collect as much light pollution as starlight. For city dwellers, a good quality refractor with its higher contrast would be a good investment. Dobsonians up to 8 inches diameter are also an excellent all-round scope for planets and deep space, while still being relatively easy to transport. Observing planets and the moon is possible even in the city, as they are bright enough to shine through the light pollution.

Of course, you'll need a clear night to begin with, and if you live in a polluted environment you'll want to get yourself and your telescope out under some clear country skies to have any chance of tracking down stars, nebulae, clusters and galaxies. There's a good probability you'll only buy one telescope in your life, so do some research, learn the sky first and choose quality.

Steady, steady

Rock-solid mountings are critical to the performance of any telescope. Wooden decks are out of the question, as are most high-rise buildings due to vibration and movement. Solid ground is really the only place for a telescope. Professional elevated telescopes use solid piers and sophisticated engineering to firmly attach to the Earth.

Seriously big telescopes

The bigger the telescope, the more light it collects. The more light it collects, the brighter the images we receive and the fainter the objects we can see. With telescopes, bigger is better!

Light is literally information. Analysing light from individual stars gives us a window into the workings of the universe. Spectrometry is a fascinating branch of astronomy where the light collected from stars is split into its spectrum with a spectrometer to reveal the processes and chemical elements present within stars.

In its simplest form a spectrometer shines starlight through a prism or crystal, splitting the light into a rainbow. Analysing a rainbow spectrum with its unique series of absorption and emission lines tells us the nature of that particular star. Spectrometry has shown us that all of the chemical building blocks of existence we know about on Earth are totally abundant throughout the universe.

To get the best value out of the biggest telescopes we put them on the world's highest mountains, so that they get to look through less of the Earth's atmosphere. Mauna Kea in Hawaii and the mountains of Chile are the favoured sites for telescopes 10 m plus in diameter and costing hundreds of millions of dollars.

The Hubble Space Telescope

Going one step better than giant telescopes on Earth is Hubble, a bus-sized telescope in orbit 400 km above the Earth. Hubble is literally the best set of eyes humanity has ever had. We get to see the stars from outside the fuzzy blanket of the atmosphere where there is no twinkle or scintillation or other distortion.

The views are superb and everything the Hubble Space Telescope does is in the public domain, which means all of its high-resolution images are freely available on the internet for anyone to use. Hubble is one of humanity's greatest assets.

What are you going to see?

My favourite stargazing pastime is to simply go outside, lie on my back and look up at the stars with my naked eye. I try to forget everything I think I might know about the cosmos and allow the stars to pour their wisdom into me.

Binoculars are my next favourite stargazing tool, thanks to their wide-angle views, which are still far narrower than the lie-on-your-back-on-the-trampoline view. You're lucky if you fit all of the Southern Cross in the field of view of a pair of binoculars!

Learning the sky is a bit like becoming familiar with a new town or meeting a new bunch of people at work or school. On the first day, you have no idea where you're going or what the stars are called. By the second day, you at least know the main street and how to get where you're going, and in a few days you can find your way all over town.

Of course, nowadays you can get your dash-mounted GPS to tell you where to go, just like you can get a computer-driven GOTO telescope that slews around and finds things for you in the sky. This technology is amazing, with stored databases of thousands of stars, planets and deep space objects.

The technology extends to handheld devices and smart-phones that allow you to point at a star and have all of the information about that star come up on screen. You can enter a name and little arrows will guide you to a particular star or planet. While all of these tools are amazing and useful, none of them replaces your own experience, which, once gained, is yours forever.

Star wheels

Begin your explorations with a simple star chart. My favourite is a cardboard or plastic star wheel or planisphere, which is highly effective at letting you know what is in the sky on a particular day at a particular time. Smartphone applications with star maps that move on the screen as you move them around the sky work very well too.

To start off, find something in the sky that you might already know, like Orion's Belt, which can be seen from everywhere on the planet. If you can't find it, it might be below the horizon. Look for Scorpio instead. The Big Dipper or Plough is pretty easy in the northern hemisphere and the Southern Cross is well known in the southern hemisphere. Next, find them on your star wheel or your smartphone and you have your starting point. Line up your star wheel with what you see in the sky and then start adding in stars by hopping across the sky from star to star. Add in a couple of new stars every night and you will be well and truly on your way to navigating your way around the sky.

Watch the sky for an hour or so and it will be pretty obvious there is some movement going on. That's when you start to see the Earth turning. If you come back three months later and look for the same stars, they will either be gone below the horizon or have shifted so far into a different part of the sky that you may have trouble finding them at all. This is the Earth going around the Sun! Our view of the sky changes week by week and month by month in a pattern that is repeated year after year.

Deep space

Telescopes are great for putting things into perspective. That fuzzy little patch of light in the telescope means very little unless you realise it's an entire Milky Way-sized galaxy 30 million light years away. Star clusters a few hundred or a few thousand light years away are also spectacular in their family groupings. Zooming in on a star and discovering that it's a double can have you wondering about what you would see if you lived on a planet with two suns in its sky.

The stars themselves are so far away that they'll only ever be pinpoints of light in the eyepiece – and the better the telescope, the sharper the pinpoints. The planets are close enough that they'll show as distinct discs, with some of them truly spectacular and others offering little detail at all.

Your eyes

When all is said and done, the only tools you really need for astronomy are your eyes. Get out there and look up – or out! If you walk straight out of a well-lit house, it's possible you won't be that impressed. Next time stay out there for 20 minutes and your eyes will adjust. Your pupils will dilate and let in the ample starlight generated by the hundreds of billions of suns that are floating in space all around us.

Chapter 24

Are There Any Questions?

'Are extra-terrestrials *real*?'

In 25 years of hosting star tours I've been asked all kinds of questions by all manner of people from all over the world. More often than not, the age-old query about whether human beings are alone in the universe, or if there is extra-terrestrial life 'out there' somewhere, is posed by a child.

It's a great question and one that all of us humans on Earth wonder about. Are we alone in the universe? I think it is the same question as 'Where have we come from?' 'Where are we going?' 'What are we here for?'

Rather than searching for answers from outside, I encourage them to look around at the stars and recognise that all of the stars are other worlds that are made of the same stuff we are. I know from my own wonderings that I'd be surprised to find that there wasn't life absolutely everywhere. However, that's just what I think. It doesn't make it right.

Even if we put some numbers on it, our home galaxy, the Milky Way, has something like 200 billion stars in it and every one of those is probably surrounded by its own family of planets. If there was an average of, say, five planets per star (ours has eight or nine, depending on who you ask), that means there could be around one trillion planets *in our own galaxy* that may or may not be home to life of some sort.

Current conservative estimates that there are around 100 billion galaxies in the universe – and I'd suggest that an infinite number of galaxies would be closer to the mark – stretch the number of planets almost beyond comprehension to 100 billion trillion. If 0.0000000001% of those planets was home to just one form of life, we'd be talking about one trillion possibilities for life in the rest of the universe.

We haven't discovered any other life yet, but let's face it, we've only set foot on the moon! And although we still have machines scratching around in the Martian dirt looking for signs of life, there are 100 billion trillion or so planets to go!

I'm also often asked if I've ever seen a UFO during my many years living beneath a canopy of stars. The answer is no. There have, however, been plenty of people who've solemnly told me about their encounters. I can explain nine out of 10 UFO stories I hear, but occasionally one comes along that stands the hair up on the back of my neck. As I say, I haven't seen any myself, so they remain just stories that I've heard. Not that part of me isn't open to the idea. In fact, I've stood on the top of desert mesas in the middle of the night and howled out at the top of my voice for extra-terrestrials to come and take me for a ride. So far I've had no response. If they have heard me, they've chosen to completely ignore me.

Are black holes real?

Everybody wants to know about black holes. They really have captured the imaginations of young and old alike. What are they and do they really exist? In 2019, people started asking me, 'Were you excited to finally see a black hole?'

In April 2019, a group of international astronomers produced the first-ever image of a black hole, in a super-giant galaxy called M87 in the constellation of Virgo.

A black hole is a theoretical object in space that has enormous mass crammed into a very small area. The thinking is that all that mass generates such colossal gravitational force that nothing can escape a black hole, not even light.

It seems fairly obvious that if there's a major source of energy in the centre of our solar system – our Sun – and our galaxy is made up of some 200 billion suns, there's some sort of energy source holding that together too.

Until recently no-one had ever claimed to have seen a black hole, so when international astronomers told the world they had photographed one, I was as intrigued as anyone else. When I saw the photo, however, I wasn't exactly blown away. It looked to me like a fuzzy smudge that could have been anything.

I'll keep watching when it comes to black holes. I'm sure there is something there, holding the galaxy together, and I'm sure it's our job as humanity on planet Earth to do our best to find out. That's why I'm grateful we've got people like Professor Brian Cox to explore the theoretical possibilities. I'm the practical guy, though, so when black holes move out of the realm of theory, I'll be right there.

Does the universe have an end?

Like the question about off-planet life, this is one the kids tend to ask. And I think that's great, because adults have long since given up trying to answer the hard questions, yet it's the same question we've all pondered at some time in our life.

I simply tell kids we just don't have a clue if or where the universe ends, and that's what makes it still one of the most important questions to keep asking! If you're a kid reading this, remember that *you* might be the person who finally comes up with the answer.

Is it true we're made of stardust?

Yes, human beings – as with everything on Earth – are made of exactly the same stuff as the stars. The periodic table of elements provides the universal building blocks of everything that exists on the physical plane. You are as much an assortment of stardust as anything else.

What is space?

What we think of as empty space also contains a certain variable density of atoms per cubic light year. This density varies from within the plane of the solar system to outside of it. It varies again in between the stars and it varies locally depending on which part of space the solar system is travelling through.

When we start to think about 'space' itself – the very fabric that everything is embedded 'in' – there is a lot of cause for wonder. Is it a 'some-thing' or is it a 'no-thing'? We can wonder about that for eternity and I suggest that you do.

If an asteroid was to hit Earth, would we see it coming?

Hollywood has made a couple of movies about the countdown to this kind of scenario in *Armageddon* and *Deep Impact*. Well, in the non-Hollywood version, if a giant comet or asteroid was going to wipe us out, we'd be unlikely to see it coming.

A lot of the objects we discover that come close to us are detected a couple of days after they've gone past! Our way of finding these near-Earth objects (or NEOs) is by looking for their sideways movement amongst the stars. There are telescopes around the world – including at Siding Spring in NSW – that are dedicated to doing just that.

That's all well and good for detecting objects that are moving at any kind of tangent, or sideways, to Earth. If something is coming directly towards us, however, it's not going to be moving amongst the stars from our perspective – it's going to be a pinpoint which would be extremely hard to identify as a moving object. It would simply keep coming towards us, getting bigger and brighter and bigger and brighter until . . .!

Wow! What a way to go!

Our planet has already been impacted many times by meteorites that were large enough to cause devastation across the world. There are dozens upon dozens of enormous craters around the globe that tell the story of these impacts (but say nothing of the monster collisions that no doubt occurred in the oceans).

In fact, I've stood on the rim of a crater about 600 km east of my home at a place called Wolfe Creek, on the other side of the Great Sandy Desert from Broome. It's 875 m in diameter and 60 m from the top of the rim to the crater floor.

It's an amazing place to walk around inside of, imagining what scientists estimate as a 15 m across and 17,000 tonnes visitor from space vaporising itself and leaving such a massive hole. Trees grow in the bottom on a mud pan with deep cracks, making you wonder how far down they go through the wind-blown sand and mud that has partially filled the crater over the past 120,000 years.

Compare that with the largest impact site on Earth, the Vredefort crater in South Africa. That crater is 300 km in diameter and was caused by an asteroid 10 to 15 km in diameter more than two billion years ago. If we get hit by one of those, then it's curtains for the human race – including Hollywood.

Are shooting stars really, y'know, shooting stars?

We call them shooting stars or falling stars because that's what they look like. However, they are not stars at all and they are not dying. They are much closer to us than the stars. In fact, what you're seeing is a little (or sometimes bigger) piece of rock entering the Earth's atmosphere and going so fast and generating so much friction that, at about 100 km up, it burns up as it comes through the air.

The correct name for a shooting star or falling star is a meteor. A meteor begins life as a meteoroid, a small piece of rock happily travelling through space minding its own business. If it happens to try and travel through the same piece of space that our planet happens to be travelling through, the meteoroid starts to burn up as it enters the atmosphere and becomes a meteor.

If anything is left over from the burning up and reaches the ground, that's what we call a meteorite. So it goes from meteoroid to meteor to meteorite. Now, not only do these rocks burn up as they come through the atmosphere, they also strip the electrons off the air molecules that they pass through, in a process called ionisation. Another word we can use for ionisation is electrification. Meteors are streaking through our sky and making electric arcs in the atmosphere at about 100 km above our heads. And this is what we see.

You can call them shooting stars – I don't have a problem with that. But just remember what they actually are and I'm sure you'll see them as just as amazing. There really is something special about them, as if they're bringing us a message. They always seem to give you a lift and the big ones seem to stay with you forever. I always make sure I see one before I fall asleep under the stars.

Are meteors the same as comets?

It's easy in our imagination to mix up comets and meteors. After all, they are both very special things in our skies. However, they are very different things that do happen to have a distinct relationship with each other. While a meteor is usually no bigger than a pinhead, a comet is typically 10 km across with a coma and tail potentially bigger than the Sun. While a meteor streaks through the sky on the spot while you are watching, a comet appears to stay relatively still in the sky and moves a little (or a lot) from one night to the next, changing size and shape and brightness as it goes. While a meteor is usually about 100 km over your head when you see it entering the Earth's atmosphere to burn up, a comet can be hundreds of millions of km away and still be visible, because the Sun is shining on it and lighting it up.

Meteors are, however, associated with comets in that many of them are small pieces of comets, debris left behind in the highly elongated orbits of the comets that spread out to make a stream of debris orbiting the Sun. With the orbit of a comet often crossing the orbit of the Earth, we pass through many of these streams of debris during the course of our year-long journey around the Sun. As we do, bits of rock and dust and gravel enter the Earth's atmosphere and we see them as shooting stars, or meteors. Many different comets have left many different streams of debris producing meteors with many different qualities of size, speed, colour, intensity and quantity. You'll find a list of meteor showers in any astronomy yearbook so that you can plan your meteor watching all throughout the year.

One of my favourite meteor showers are the Leonids that streak through our skies at 72 km/s every 17, 18 and 19 of November. The many different meteor showers that happen throughout the year have different speeds that vary from 11 to 72 km/s. The Leonids are the fastest of them all. They leave big long smoke trails across the sky from one side to the other. You can usually get at least a few of these per hour in most years, although in some years it can be many more.

In 1999, I was camping out with a group of ESA scientists who had come out from Europe for the event and we saw 1000 shooting stars between 2am and dawn. That's a lot of wishes on one of the most amazing nights of my life. What are you doing this November 17, 18 and 19 in the early hours of the morning? See you there!

Chapter 25

Earth-turning Consciousness

No matter how many times I do it – and it's into the thousands now – I'm still filled with awe and reverence when I stretch out beneath the stars and wait for a meteor to draw a sparkling line across the night sky and finish off my day. It's not just my cue to go to sleep and let Earth carry me through the darkness once again, it's a reminder the cosmos is as intriguing for its order and perfection as it is for its seeming chaos.

We don't know how the universe came to be, whether it even *had* a beginning, or if it will someday come to an end. We don't know whether it is all under control and if there is a plan. Somehow I think there must be and that each one of us is responsible for doing our part in it, just as the planets and the stars in their interactions with each other are doing theirs. In watching what I've learned to watch so far, it seems to me that there's an obvious intelligence and order to the universe and that it will all unfold for us in its own good time.

Although meteoroids and particles randomly zip in every direction through space, the seemingly meticulous arrangement of billions of heavenly bodies – like the moon's near-perfect relationship to the Earth and the elegant layout of countless far-flung spiral galaxies – does give one cause for thought.

I'm also struck by the amazing coincidences that have adorned the map of humankind's exploration of space so far. It's remarkable that all of the outer planets, bar Pluto, lined up exactly as they needed to be at the exact time in history when humans had conjured the technology to fire probes at them. It's almost as if the solar system had somehow been geared to wait for us. Or was it grooming us for the task at hand?

This odyssey began with Pioneer 10's flyby of Jupiter in 1973, followed by Pioneer 11 in 1974. Thanks to the alignment of the planets at that time, Pioneer 11 was able to continue on towards Saturn.

The Pioneers paved the way for the Voyagers in 1977. Voyager 1 took off in September that year and whizzed past Jupiter in March 1979, before scooting off to check Saturn out in November 1980. Pioneer 11 had revealed enough of Saturn's moon, Titan, for NASA to want a better look, so Voyager 1 was sent to check it out.

Voyager 2 was launched on a slower trajectory in August 1977. It visited Jupiter in July 1979, Saturn in August 1981 and kept racing through the plane of the solar system to Uranus in January 1986 and Neptune in August 1989. It then continued well out beyond the orbit of Pluto, analysing its environment as it probed the edge of the heliosphere – the Sun's influence in space.

As you read this, Voyager 2 is barrelling towards Sirius some 8.6 light years away and it's likely to pass it – the brightest star in our skies – in about 296,000 years, before continuing on to who knows where. Maybe it will eventually slam into a big brick wall.

Should a life form intercept this deep-space extension of the human race, we've drawn a few pictures on it to let them know where we come from. We've also stashed a phonograph record on board so they can listen to Chuck Berry and Mozart, if they ever figure out how to play it.

Those early missions were monumental and they super-charged our knowledge and understanding of the cosmos. Incredibly, though, they have all taken place within one lifetime – this lifetime! Most amazing of all to me, however, is the fact we simply could not replicate those voyages of discovery today, because the planets are not aligned in a way that would allow it.

Jupiter and Saturn pass each other in the sky once every 20 years, but to bring Uranus and Neptune into alignment with them at the same time happens only every 175 years. They *were* lined up back in the 1970s, just when we so happened to have the right spaceships ready. If that's not the most insane coincidence in human history, then I don't know what is.

Thinking about these things has also made me cognisant of the galactic signposts that have marked my own journey aboard this beautiful, spinning rock. Consider the crash between the comet Shoemaker-Levy 9 and Jupiter in 1994 – yet another spectacular alignment of factors that allowed us to witness the first-ever cosmic collision in known human history.

For starters, the Hubble Space Telescope had been launched in 1990 and, after a few teething problems, was finally fully operational by 1993 – the same year Eugene Shoemaker and

David Levy discovered their comet and realised a head-on encounter with Jupiter was imminent.

Not only was the Hubble in position in low Earth orbit the following year to observe the results of the collisions, the Galileo spaceship just happened to have arrived slightly around the back side of Jupiter in time to capture the actual locations where the huge chunks of the disintegrating comet ploughed into Jupiter's surface.

Meanwhile on Earth, I just so happened to have a new 10-inch Dobsonian telescope, which gifted me a pretty good view of celestial history in the making too.

In the absence of any other explanations, science tells us to regard these confluences of events as coincidences, but having watched the sky and our place in it so closely for so long, and having noticed the flow and rhythms inherent in the myriad cycles around us, I choose to leave the door ajar and see 'coincidences' as perhaps something more than uncanny flukes.

When I witnessed Halley's Comet in 1986, I strongly felt it was a messenger bringing something new into our inner solar system, and for me that's exactly what it was. Ten years later – not long after I tentatively started presenting the stars to the public in an attempt to make it my livelihood – the dazzling Comet Hyakutake showed up in our skies. I instantly took it as a sign that starting Astro Tours was the right move in my life. I still do.

By slowly and irreversibly tuning in to the movements of Earth as it travels through space, I feel 100% certain I'm exactly where I'm supposed to be and doing exactly what I'm supposed to be doing. There's no other way it can be. Waking up to the reality that we are on a rocky spaceship that is hurtling through space in a number of different directions at once has made a huge difference in my outlook and how I interact with the world around me.

It sounds a little crazy and I don't know how to express it other than to say I *know* the direction this planet is moving in. I *know* where it's going and I *know* I'm just along for that ride. I can choose to either go along with it or resist it. I chose to consciously be part of the journey long ago. It's what I call 'flow'.

Over the last quarter of a century, I've presented the sky to well over 100,000 people in live performance, and a lot of them come back again and again over the years and tell me about the same changes related to flow in their own lives. Since 2017 I've shared with millions of people through my television shows, and I get social media messages and emails from all over the world saying the same thing.

People share stories with me about finding flow and direction as a direct result of knowing where and how the Earth is heading on its journey. This feedback is beautiful and it confirms my own experience. Personally, it has led me to a place of peace and belonging. The Earth, the planets, the Sun and all of the stars in all of existence are exactly in their places too. It makes sense that this is true for each and every one of us.

I see the only way to make this concept a reality in your own life is by getting out under the stars. Of course, I live in a pretty remote part of the world and I spend a lot of my time looking at and thinking about space, so my embrace of this world view – which I call 'Earth-turning consciousness' – has been a pretty logical and simple progression. However, I recognise that millions of people, perhaps billions, live hectic lives brimming with earthly distractions that prevent them from taking the time to look at the bigger picture and the longer journey.

Yet there's something even more fundamental about the human race that stops so many of us from tuning in to the movement of the planet through the cosmos: and that is the way we speak.

Even though you know the Earth is turning, I'd be willing to bet that at least once this week you've flatly denied this is so. What? You wouldn't do that? Well, I'm sure you wouldn't *knowingly* deny the Earth is spinning, but if you said something like 'Let's go and watch the Sun go down' or 'Let's go and watch the moon rise', you would have done just that.

We effectively blind ourselves when we talk about the Sun 'rising' and the moon 'setting', or when we make reference to 'up' or 'down' in relation to stars and planets. The Sun doesn't go down at all. The Earth's rotation just makes it appear that way.

Yes, I have used what I call 'flat-Earth language' in the preceding pages of this book, but only because I didn't want to mess with your head straight out of the gate – I thought I'd save that for the end. Still, identifying that we have a problem with flat-Earth language is one thing, coming up with a way to fix it is quite another.

I challenge you to have a crack right now at finding a succinct way to describe the Earth turning away from the Sun at the end of the day, and see if you can conjure anything as simple as 'sunset' – which you're likely to now see as a not quite right way of talking about it. In other words, try to come up with a factually and scientifically correct way of saying, 'Let's go and watch the sunset' – just without using the word 'sunset' . . .

How'd you go? Not so great, huh? It's not because you don't have great command of language – you do! The reason you have difficulty expressing the simple action of the Earth turning is because there is no single, definitive word that refers to it. Not one! In any language!

While we use single words to describe 'day', 'night', 'morning', 'afternoon' and so on, there is no word that describes the humungous process that causes it all. The only way we can currently communicate the Earth's rotation is by using comparatively long-winded sentences that require us to think about what we're saying and acknowledge what's really going on. I've found that waving my arms around to make myself understood seems to help too.

I suppose this is because the languages of the human race developed over millennia. Our linguistic roots took hold long before the scientific revolution – led by the likes of Copernicus, Kepler, Brahe, Galileo et al – gave us the intellectual knowledge that the Earth is turning. Remember that some did so at great personal risk, like poor ol' Giordano Bruno, who was burnt alive for daring to propose the stars are suns.

Those pioneers were willing to buck the system, to think for themselves and acknowledge the realities that they took the time to observe and ponder. Isn't it about time we honoured them by taking time in our own lives to translate our personal intellectual knowledge of the Earth turning into experience? If enough of us spend a greater proportion of our days living with an Earth-turning consciousness, perhaps one of us will come up with the right language to maintain this higher state of awareness.

As you can imagine, I've given a lot of thought to 'Earth-turning language' over the years. And I've come up with something that works. Now, I'm not going to claim that it's particularly easy, or neat and tidy, or that it's even going to catch on. What I am going to do, though, is try it out on you!

Instead of saying, 'Let's go and watch the Sun go down,' you'd have to say something like 'Let's-go-and-watch-Earth-turn-away-from-the-part-of-the-sky-that-the-Sun-is-in.'

Is that romantic or what? If you say, 'Hey, darling, let's go and watch the Earth turn away from the part of the sky that the Sun is in,' what is your date going to do? They're probably going to go and watch the 'sunset' with someone else!

It is a simple statement of fact. When we look at the 'sunset' we're purely and simply watching the part of the Earth that we are on turn away from the part of the sky that the Sun is in. I don't know of a shorter way of telling the truth about what's happening when we watch a sunset. It doesn't sound very poetic or romantic. To me it sounds like a yawning gap in our language.

Words are powerful things. In many ways they shape our reality. If we say, 'Let's go and watch the Sun go down,' for us the Sun is going down! But if we say, 'Let's go and watch the Earth turn away from the part of the sky that the Sun is in,' watch what happens in your life!

Actually, a few years ago now, I had this very conversation with a guy out surfing on the Bukit Peninsula in Bali. He reminded me that when we were in school, in relation to words, we were all taught to spell. What's another meaning of the word 'spell'? We cast spells on ourselves with the words we use.

If we change the words, we can change the spells. And if we've been kidding ourselves about such a fundamental truth as the Earth turning, for so long and so effectively, what else have we been kidding ourselves about? What other illusions are we living under that might just happen to fall away if we manage to crack this flat-Earth thing that we've all been so actively participating in? Might be worth giving it a go.

So now that you've got a few ideas about where you are in the universe and where you're headed, I'd love it if you could come up with your own words, your own way to describe what's going on when you next sit down to partake in the sacred ritual of the sunset, and let me know. If you're anything like me, your words will evolve, and your understandings will grow a little bit more every time you do it.

Anyway, why don't you sleep on it? A shooting star will be along any moment now.

Resources

Yearbooks

I've been selling astronomy yearbooks from Ken Wallace, Glenn Dawes and Peter Northfield of Quasar Publishing almost as long as they have been producing them. Written by Australians for Australians, they will let you know all of the things that change each year in the sky; like the planets! Designed for anyone who looks at the night sky, whether you are using just your eyes, a pair of binoculars or a telescope, they have something for everyone from the basic novice up to the advanced amateur astronomer. This includes those with a casual interest who might just want to know, 'What is that bright star next to the moon?'

Star charts

A Night Sky planisphere is a wonderful tool for learning your way around the sky. Turn the two wheels to match your observing date and time and it will show you what stars are in the sky. Because the stars are as regular as clockwork during the night and during the year, you only have to buy one in your whole lifetime! A planisphere makes a great companion to a yearbook.

Apps

There are any number of star apps out there. 'Quickie, they are going to put you out of business' is what everyone told me. What they have done is drum up business by spiking everyone's imagination. 'Do they really work?' is another one I get all the time. Yes, they do. They are hooked into the GPS in your phone with a database that has me reaching for them first before I ever get out my old star atlases. They are wonderful.

Websites

Again there are so many resources online that will take you down so many rabbit holes that I am only going to leave you with one. Simply search for Hubble Space Telescope and you will have access to everything that this school-bus-sized space telescope has done since the early 1990s in full resolution, for free!

Acknowledgements

Ever since the first nights out there under the stars with small groups of people a quarter of a century ago, I've been asked for a book. This has been an incredibly profound gift to me from those who have asked, as they have let me know the importance of bringing out into the open those things that I am here on Earth to share. I see clearly that each one of us has that thing that is ours to share and to contribute to make this world and this universe the amazing entity that it is.

The sky shows me this with the interconnectedness, the obvious order and intelligence displayed for us by the journey of the planets, the stars and the galaxies through space, each one contributing to the existence of the whole. That's why I am grateful to you, for doing your part, for saying your piece, for running your life in your own way, for thinking for yourself and for coming to your own conclusions. We'll all make mistakes, take wrong turns, and bump into things when we are not looking. These are the experiences that shape us, steer us and bring us always to that which is next in our life.

The steps in my own life have been punctuated with teachers and guides who have turned up at exactly the right time. I am grateful to my family, my teachers, my friends, those who have encouraged me and those who have provided me with opportunity and challenge. You have met some of them already throughout the book, a book made possible by the wonderful people at Penguin Random House, who know what they are doing when it comes to getting stories out there that benefit the world. I remember well the day I received a lovely phone call from Sophie Ambrose, asking if I would like to write a book. Sophie's enthusiasm for life and books shines through clearly in everything that she does. Thanks to Craig Henderson, who came to Broome to sample a bit of Space Gandalf time before going home with a stack of material from me to turn into a book. We quickly became firm friends. Patrick Mangan with his eagle eyes ready to pick me up on all the times I've gone off track or repeated myself has been a joy to work with. Thank you all and thanks also to your team who are at the top of their game in the world. Dr Brad E. Tucker, astrophysicist/cosmologist at Mount Stromlo Observatory, kindly agreed to cast a professional eye over my work with some wonderful and welcome suggestions.

I am grateful to every person who comes into my life on every single day, whether you are my blood family, my bush family or my cosmic family, I salute you, I smile with you, I walk with you and I fly with you as we journey together through the stars.

About the Author

Outback astronomer **Greg Quicke** runs astronomy tours from Broome and goes by the nickname 'Space Gandalf'. After appearing on the ABC's *Stargazing Live* he became an overnight celebrity. He also starred in *A Stargazer's Guide to the Cosmos* in which he took viewers on the ultimate guided tour of the southern sky, revealing unseen connections between the everyday world around us and the stars above.